MATHEMATICS FOR FISHERIES ECONOMICS

A BASIC MANUAL

MATHEMATICS FOR FISHERIES ECONOMICS

A BASIC MANUAL

– Author –

Dr. Vinod Kumar Yadav

PhD, IIT Bombay
Senior Scientist (Agricultural Statistics)
Fisheries Economics, Extension and Statistics Division
ICAR-Central Institute of Fisheries Education(CIFE),
Mumbai, Maharashtra, India

2023

Daya Publishing House®
A Division of
Astral International Pvt. Ltd.
New Delhi – 110 002

ISBN 9789359191546

Publisher's Note:

Every possible effort has been made to ensure that the information contained in this book is accurate at the time of going to press, and the publisher and author cannot accept responsibility for any errors or omissions, however caused. No responsibility for loss or damage occasioned to any person acting, or refraining from action, as a result of the material in this publication can be accepted by the editor, the publisher or the author. The Publisher is not associated with any product or vendor mentioned in the book. The contents of this work are intended to further general scientific research, understanding and discussion only. Readers should consult with a specialist where appropriate.

Every effort has been made to trace the owners of copyright material used in this book, if any. The author and the publisher will be grateful for any omission brought to their notice for acknowledgement in the future editions of the book.

Published by : **Daya Publishing House®**
A Division of
Astral International Pvt. Ltd.
– ISO 9001:2015 Certified Company –
4736/23, Ansari Road, Darya Ganj
New Delhi-110 002
Ph. 011-43549197, 23278134
e-mail: info@astralint.com
Website: www.astralint.com

Acknowledgements

The author is extremely thankful to Dr. Ravishankar C. N., Director/Vice-Chancellor, ICAR-CIFE, for providing all necessary facilities to carry out the work. Author would like to express his gratitude to Dr. Gopal Krishna, former Director, ICAR-CIFE, Mumbai for his constant support, encouragement and motivation. The author sincere gratitude to Dr. S.N. Ojha , Head, Fisheries Economics, Extension and Statistics (FEES) Division, CIFE Mumbai for his constant motivation and encouragement in every stepto accomplish this journey. The Author expresses his pleasure and thanks to all his family members who always stand on his side for all the support and encouragement provided to accomplish this journey. Finally, author would like to thank his colleagues at FEESD, his students at ICAR-CIFE and whoever in a wide range of spectrum helped him. Special thanks are to Mr. Suresh and Ms. Sneha, his MFSc-Fisheries Economics (FEC) students for their painstaking efforts in evaluating the manual and providing useful suggestions.

Dr. Vinod Kumar Yadav

Foreword

ICAR-Central Institute of Fisheries Education (CIFE) at Mumbai is a premier institute cum Deemed University, which is one of its kinds in India conducting academic programs (M.F.Sc. & Ph.D.) in eleven core disciplines of fisheries sciences. In addition, the institute also conducts basic and strategic research in frontier areas of fisheries. The institute occupies a place of pride in organising demand-driven training and educational programs, providing technical support, inputs for policy development and consultancy services to various stakeholders in fisheries, aquatic and aquaculture sectors.

The Fisheries Economics, Extension and Statistics (FEES) Division is one of the important divisions engaged in social science education, training and research oriented towards growth and development of fisheries sector. The statistics faculty of the division are engaged in teaching Mathematics, Statistical methods, GIS and Remote sensing, AI and big data Analytics, not only to Fisheries Economics and Extension students, but also to the other nine disciplines of the institute viz., Aquaculture, Fisheries Resources Management, Post-Harvest Technology, Fish Nutrition and Feed Technology, Fish Biochemistry and Physiology, Fish Genetics and Breeding, Fish Biotechnology, Aquatic Environment Management and Aquatic Animal Health.

With advances in information technology, the institute has all along been adapting itself to the current needs and methodological challenges and quality enrichment of fisheries education. The increasing application of mathematics to various branches/courses of economics in the past decade has made it necessary for economists to have an elementary knowledge of mathematics. To this end, an attempt has been made to bring out a practical manual on mathematics for catering to needs of fisheries students, most of whom having a non-mathematical background during their tertiary education.

The practical manual is written by Dr. V. K. Yadav, Senior Scientist, ICAR-CIFE, Mumbai published by our institute. The manual has been structured to solve problems on a wide range of topics such as matrix, determinants, limit & continuity, differentiation, integration, partial differentiation, etc. It is hoped that the manual brought out will serve as a useful source of knowledge to the students and also will help them in solving various types of mathematical problems. I wish to complement the authors Dr. V. K. Yadav, ICAR-CIFE of this institute for his painstaking efforts in bringing out this valuable manual for the benefit of students in fisheries economics.

Dr. Ravishankar C. N.

Director/Vice-Chancellor,
ICAR-CIFE, Mumbai

Preface

This book is an outcome of the feeling that a book on "Mathematics for Fisheries Economics: A Basic Manual" is an urgent need for post-graduation students pursuing courses in Fisheries Economics as well as the allied discipline of fisheries sciences. The students enrolled for fisheries Economics and allied discipline of fisheries science are generally from the biological background. As the mathematical base is very much required to go for different courses of Economics like Macro Economics, Microeconomics, Econometrics, International economics and Trade, GIS and Remote sensing, etc., and also for some minor and supporting courses like Fish stock assessment methods, Statistical methods, Artificial Intelligence and Big data Analytics. The book has been structured to solve problems on a wide range of topics such as matrix, determinants, limit & continuity, differentiation, integration, partial differentiation, Differential equation, finite difference, and Numerical Analysis. It is hoped that the book brought out will serve as a useful source of knowledge to the students and also will help them in solving various types of mathematical problems.

While every effort has been made to make the book useful and error-free but there is always scope for further improvement. I shall be grateful to the readers for their valuation suggestions and criticism to improve the book.

Vinod Kumar Yadav

Contents

Chapter 1

Matrix and Determinants

Introduction to Matrices

1. **Construct 2×2 matrix, $A = [a_{ij}]$, whose elements are given by:**

$$a_{ij} = \frac{(i+j)^2}{2}$$

Since it is 2 x 2 matrix

It has 2 rows and 2 columns

Let matrix be A

Where $A = \begin{bmatrix} a_{11} & a_{12} \\ a_{21} & a_{22} \end{bmatrix}$

Now it is given that

$$a_{ij} = \frac{(i+j)^2}{2}$$

a_{ij}	$i =, j =$	$a_{ij} = \frac{(i+j)^2}{2}$
a_{11}	$i = 1, j = 1$	$a_{11} = \frac{(1+1)^2}{2} = \frac{(2)^2}{2} = \frac{4}{2} = 2$

a_{12}	$i = 1, j = 2$	$a_{12} = \frac{(1+2)^2}{2} = \frac{3^2}{2} = \frac{9}{2}$
a_{21}	$i = 2, j = 1$	$a_{21} = \frac{(2+1)^2}{2} = \frac{3^2}{2} = \frac{9}{2}$
a_{22}	$i = 2, j = 2$	$a_{22} = \frac{(2+2)^2}{2} = \frac{4^2}{2} = \frac{16}{2} = 8$

Hence,

The required matrix A if

$$A = \begin{bmatrix} a_{11} & a_{12} \\ a_{21} & a_{22} \end{bmatrix} = \begin{bmatrix} 2 & \frac{9}{2} \\ \frac{9}{2} & 8 \end{bmatrix}$$

Algebra of Matrices

Addition

To add two matrices: add the numbers in the matching positions

2. $\begin{bmatrix} \mathbf{3} & \mathbf{8} \\ \mathbf{4} & \mathbf{6} \end{bmatrix} + \begin{bmatrix} \mathbf{4} & \mathbf{0} \\ \mathbf{1} & \mathbf{-9} \end{bmatrix} = ?$

Solution

$$\begin{bmatrix} 3 & 8 \\ 4 & 6 \end{bmatrix} + \begin{bmatrix} 4 & 0 \\ 1 & -9 \end{bmatrix} = \begin{bmatrix} 7 & 8 \\ 5 & -3 \end{bmatrix}$$

As

$3 + 4 = 7$

$8 + 0 = 8$

$4 + 1 = 5$

$6 - 9 = -3$

The two matrices must be the same size, i.e., the rows must match in size, and the column must match in size.

Negative

3. $-\begin{bmatrix} 2 & -4 \\ 7 & 10 \end{bmatrix} = ?$

Solution

$$-\begin{bmatrix} 2 & -4 \\ 7 & 10 \end{bmatrix} = \begin{bmatrix} -2 & 4 \\ -7 & -10 \end{bmatrix}$$

Subtraction

4. $\begin{bmatrix} 3 & 8 \\ 4 & 6 \end{bmatrix} - \begin{bmatrix} 4 & 0 \\ 1 & -9 \end{bmatrix} = ?$

Solution

To subtract two matrices: subtract the number in matching positions:

$$\begin{bmatrix} 3 & 8 \\ 4 & 6 \end{bmatrix} - \begin{bmatrix} 4 & 0 \\ 1 & -9 \end{bmatrix} = \begin{bmatrix} -1 & 8 \\ 3 & 15 \end{bmatrix}$$

Note: subtracting is actually defined as the addition of a negative matrix: $A = (-B)$

Multiply by a Constant

5. $2\ x \begin{bmatrix} 4 & 0 \\ 1 & -9 \end{bmatrix} = ?$

Solution

$$2\ x \begin{bmatrix} 4 & 0 \\ 1 & -9 \end{bmatrix} = \begin{bmatrix} 8 & 0 \\ 2 & -18 \end{bmatrix}$$

We call the constant a scalar, so officially this is called “scalar multiplication”

Multiplying by Another Matrix

6. **Multiply** $[A] = \begin{bmatrix} 1 & 2 & 3 \\ 4 & 5 & 6 \end{bmatrix}$ ***with*** $[B] = \begin{bmatrix} 7 & 8 \\ 9 & 10 \\ 11 & 12 \end{bmatrix}$

Solution

To multiply a matrix by another matrix we need to do the dot product of rows and columns.

To work out the answer for the 1st row and 1st column:

$$\begin{bmatrix} 1 & 2 & 3 \\ 4 & 5 & 6 \end{bmatrix} x \begin{bmatrix} 7 & 8 \\ 9 & 10 \\ 11 & 12 \end{bmatrix} = \begin{bmatrix} 58 & \\ & \end{bmatrix}$$

The "Dot product" is where we multiply matching members, then sum up:

(1, 2, 3) • (7, 9, 11) = 1×7 + 2×9 + 3×11 = 58

We match the 1st members (1 and 7), multiply them, likewise for the 2nd members (2 and 9) and the 3rd members (3 and 11), and finally sum them up.

Want to see another example? Here it is for the 1st row and 2nd column:

$$\begin{bmatrix} 1 & 2 & 3 \\ 4 & 5 & 6 \end{bmatrix} x \begin{bmatrix} 7 & 8 \\ 9 & 10 \\ 11 & 12 \end{bmatrix} = \begin{bmatrix} 58 & 64 \\ & \end{bmatrix}$$

(1, 2, 3) • (8, 10, 12) = 1×8 + 2×10 + 3×12 = 64

We can do the same thing for the 2nd row and 1st column:

(4, 5, 6) • (7, 9, 11) = 4×7 + 5×9 + 6×11 = 139

And for the 2nd row and 2nd column:

(4, 5, 6) • (8, 10, 12) = 4×8 + 5×10 + 6×12 = 154

And we get

$$\begin{bmatrix} 1 & 2 & 3 \\ 4 & 5 & 6 \end{bmatrix} \times \begin{bmatrix} 7 & 8 \\ 9 & 10 \\ 11 & 12 \end{bmatrix} = \begin{bmatrix} 58 & 64 \\ 139 & 154 \end{bmatrix}$$

In General:

To multiply an m×n matrix by an n×p matrix, the "n" must be the same, and the result is an m×p matrix.

$m * n \ x \ n * p \rightarrow m * p$

Where m is no. of rows in first matrix, n is no of columns in first matrices and no of rows in second matrix and p is the no of columns in second matrices.

7. Multiply $[A] = [1 \quad 2 \quad 3] \ with \ [B] = \begin{bmatrix} 4 \\ 5 \\ 6 \end{bmatrix}$

Solution

$$[1 \quad 2 \quad 3]\begin{bmatrix} 4 \\ 5 \\ 6 \end{bmatrix} = [1 \times 4 + 2 \times 5 + 3 \times 6] = [32]$$

8. If A = $\begin{bmatrix} 1 & 1 & -1 \\ 2 & -3 & 4 \\ 3 & -2 & 3 \end{bmatrix}$**, B=** $\begin{bmatrix} -1 & -2 & -1 \\ 6 & 12 & 6 \\ 5 & 10 & 4 \end{bmatrix}$ **& C=** $\begin{bmatrix} -1 & -1 & 1 \\ 2 & 2 & -2 \\ -3 & -3 & 3 \end{bmatrix}$**. Find**

AB, BA, AC & CA

Solution

$$AB = \begin{bmatrix} 1 & 1 & -1 \\ 2 & -3 & 4 \\ 3 & -2 & 3 \end{bmatrix} \times \begin{bmatrix} -1 & -2 & -1 \\ 6 & 12 & 6 \\ 5 & 10 & 4 \end{bmatrix}$$

$$= \begin{bmatrix} -1+6-5 & -2+12-10 & -1+6-4 \\ -2-18+20 & -4-36+40 & -2-18+16 \\ -3-12+15 & -6-24+30 & -3-12+12 \end{bmatrix} = \begin{bmatrix} 0 & 0 & 1 \\ 0 & 0 & -4 \\ 0 & 0 & -3 \end{bmatrix}$$

$$BA = \begin{bmatrix} -1 & -2 & -1 \\ 6 & 12 & 6 \\ 5 & 10 & 4 \end{bmatrix} \times \begin{bmatrix} 1 & 1 & -1 \\ 2 & -3 & 4 \\ 3 & -2 & 3 \end{bmatrix}$$

$$= \begin{bmatrix} -1-4-3 & -1+6+2 & 1-8-3 \\ 6+24+18 & 6-36-12 & -6+48+18 \\ 5+20+12 & 5-30-8 & -5+40+12 \end{bmatrix}$$

$$\begin{bmatrix} -8 & 7 & -10 \\ 48 & -42 & 60 \\ 37 & -33 & 47 \end{bmatrix}$$

$$AC = \begin{bmatrix} 1 & 1 & -1 \\ 2 & -3 & 4 \\ 3 & -2 & 3 \end{bmatrix} \times \begin{bmatrix} -1 & -1 & 1 \\ 2 & 2 & -2 \\ -3 & -3 & 3 \end{bmatrix} = \begin{bmatrix} 4 & 4 & -4 \\ -20 & -20 & 20 \\ -16 & -16 & 16 \end{bmatrix}$$

$$CA = \begin{bmatrix} -1 & -1 & 1 \\ 2 & 2 & -2 \\ -3 & -3 & 3 \end{bmatrix} \times \begin{bmatrix} 1 & 1 & -1 \\ 2 & -3 & 4 \\ 3 & -2 & 3 \end{bmatrix} = \begin{bmatrix} 0 & 0 & 0 \\ 0 & 0 & 0 \\ 0 & 0 & 0 \end{bmatrix}$$

9. If $A = \begin{bmatrix} 1 & 2 & 3 \\ 4 & 5 & 6 \\ 6 & -3 & 0 \end{bmatrix}$ **and** $B = \begin{bmatrix} 6 & 3 & 1 \\ 2 & 4 & -8 \\ 3 & -6 & 1 \end{bmatrix}$**, find AB & BA**

Solution

$$AB = \begin{bmatrix} 1 & 2 & 3 \\ 4 & 5 & 6 \\ 6 & -3 & 0 \end{bmatrix} \times \begin{bmatrix} 6 & 3 & 1 \\ 2 & 4 & -8 \\ 3 & -6 & 1 \end{bmatrix}$$

$$= \begin{bmatrix} 6+4+9 & 3+8-18 & 1-16+3 \\ 24+10+18 & 12+20-36 & 4-40+6 \\ 36-6+0 & 18-12+0 & 6+24+0 \end{bmatrix}$$

$$= \begin{bmatrix} 19 & -7 & -12 \\ 52 & -4 & -30 \\ 30 & 6 & 30 \end{bmatrix} \begin{bmatrix} 19 & -7 & -12 \\ 52 & -4 & -30 \\ 30 & 6 & 30 \end{bmatrix}$$

$$BA = \begin{bmatrix} 6 & 3 & 1 \\ 2 & 4 & -8 \\ 3 & -6 & 1 \end{bmatrix} \times \begin{bmatrix} 1 & 2 & 3 \\ 4 & 5 & 6 \\ 6 & -3 & 0 \end{bmatrix}$$

$$= \begin{bmatrix} 6+12+6 & 12+15-3 & 18+18+0 \\ 2+16-48 & 4+20+24 & 6+24+0 \\ 3-24+6 & 6-30-3 & 9-36+0 \end{bmatrix}$$

$$= \begin{bmatrix} 24 & 24 & 36 \\ -30 & 48 & 30 \\ -15 & -27 & -27 \end{bmatrix}$$

10. If $\begin{bmatrix} 2 & 3 & 4 & 5 \end{bmatrix}$ **and B=** $\begin{bmatrix} 1 & 2 & 3 & 4 \end{bmatrix}$**, find AB**

Solution

A and B both are in order 1×4. So, here the multiplication is not possible as the column of A is not equal with row of B

11. If A= $\begin{bmatrix} 0 & 1 & 2 \\ 1 & 2 & 0 \end{bmatrix}$ **and B=** $\begin{bmatrix} 0 & 1 \\ -1 & 0 \\ 0 & 1 \end{bmatrix}$**, find AB and BA**

Solution

$$AB = \begin{bmatrix} 0 & 1 & 2 \\ 1 & 2 & 0 \end{bmatrix} \times \begin{bmatrix} 0 & 1 \\ -1 & 0 \\ 0 & 1 \end{bmatrix} = \begin{bmatrix} -1 & 2 \\ 2 & 1 \end{bmatrix}$$

$$BA = \begin{bmatrix} 0 & 1 \\ -1 & 0 \\ 0 & 1 \end{bmatrix} \times \begin{bmatrix} 0 & 1 & 2 \\ 1 & 2 & 0 \end{bmatrix} = \begin{bmatrix} 1 & 2 & 0 \\ 0 & -1 & -2 \\ 1 & 2 & 0 \end{bmatrix}$$

12. If A= $\begin{bmatrix} 1 & 2 \\ 2 & 3 \\ 3 & 4 \end{bmatrix}$ **and B=** $\begin{bmatrix} 3 & -2 & 1 \\ 0 & 1 & -2 \\ 3 & -4 & 5 \end{bmatrix}$**, find AB and BA**

Solution

AB is not possible. As, A is in order 3×2 and B is in order 3×3. So, here the multiplication is not possible as the column of A is not equal with row of B

$$BA = \begin{bmatrix} 3 & -2 & 1 \\ 0 & 1 & -2 \\ 3 & -4 & 5 \end{bmatrix} \times \begin{bmatrix} 1 & 2 \\ 2 & 3 \\ 3 & 4 \end{bmatrix} = \begin{bmatrix} 2 & 4 \\ -4 & -5 \\ 10 & 14 \end{bmatrix}$$

13. If A= $\begin{bmatrix} 1 & 2 & 3 \\ 4 & 5 & 6 \\ 7 & 8 & 9 \end{bmatrix}$**, B=** $\begin{bmatrix} 6 & 4 & 7 \\ 2 & 4 & 8 \\ 3 & 6 & 8 \end{bmatrix}$ **and C=** $\begin{bmatrix} 5 & 6 \\ 7 & 8 \\ 8 & -7 \end{bmatrix}$**, find (AB)C**

Solution

$$AB = \begin{bmatrix} 1 & 2 & 3 \\ 4 & 5 & 6 \\ 7 & 8 & 9 \end{bmatrix} \times \begin{bmatrix} 6 & 4 & 7 \\ 2 & 4 & 8 \\ 3 & 6 & 8 \end{bmatrix} = \begin{bmatrix} 19 & 30 & 37 \\ 52 & 72 & 116 \\ 85 & 114 & 158 \end{bmatrix}$$

$$(AB)C = \begin{bmatrix} 19 & 30 & 37 \\ 52 & 72 & 116 \\ 85 & 114 & 158 \end{bmatrix} \times \begin{bmatrix} 5 & 6 \\ 7 & 8 \\ 8 & -7 \end{bmatrix} = \begin{bmatrix} 681 & 25 \\ 1692 & 76 \\ 85 & 127 \end{bmatrix}$$

Adjoint of a Square Matrix

For every square matrix A of order n, A (adj A) = (adj A) A = |A| In.

14. If A = $\begin{bmatrix} 8 & -6 & 2 \\ -6 & 7 & -4 \\ 2 & -4 & 3 \end{bmatrix}$**, verify that A (adj A) = (adj A) A = | A | I_3.**

Solution

We find that

$$|A| = \begin{vmatrix} 8 & -6 & 2 \\ -6 & 7 & -4 \\ 2 & -4 & 3 \end{vmatrix} = 8(21-16) + 6(-18+8) + 2(24-14)$$

$$= 40 - 60 + 2 = 0$$

By definition of adjoint, we get

Adj A

$$\begin{bmatrix} (21-16) & -(-18+8) & (24-14) \\ -(-18+8) & (24-4) & -(-32+12) \\ (24-14) & -(-32+12) & (56+36) \end{bmatrix}^T = \begin{bmatrix} 5 & 10 & 10 \\ 10 & 20 & 2010 \\ 10 & 20 & 20 \end{bmatrix}$$

$$\text{A(adj A)} = \begin{bmatrix} 8 & -6 & 2 \\ -6 & 7 & -4 \\ 2 & -4 & 3 \end{bmatrix} \times \begin{bmatrix} 5 & 10 & 10 \\ 10 & 20 & 2010 \\ 10 & 20 & 20 \end{bmatrix}$$

$$= \begin{bmatrix} 40-60+20 & 80-120+40 & 80-120+40 \\ -30+70-40 & -60+140-80 & -60+140-80 \\ 10-40+30 & 20-80+60 & 20-80+60 \end{bmatrix}$$

$$= \begin{bmatrix} 0 & 0 & 0 \\ 0 & 0 & 0 \\ 0 & 0 & 0 \end{bmatrix} = 0I_3 = |A|I_3,$$

Similarly, we get

$$(adj\ A)A = \begin{bmatrix} 5 & 10 & 10 \\ 10 & 20 & 20 \\ 10 & 20 & 20 \end{bmatrix} \times \begin{bmatrix} 8 & -6 & 2 \\ -6 & 7 & -4 \\ 2 & -4 & 3 \end{bmatrix}$$

$$= \begin{bmatrix} 40-60+20 & -30+70-40 & 10-40+30 \\ 80-120+40 & -60+140-80 & 20-80+60 \\ 80-120+40 & -60+140-80 & 20-80+60 \end{bmatrix}$$

$$= \begin{bmatrix} 0 & 0 & 0 \\ 0 & 0 & 0 \\ 0 & 0 & 0 \end{bmatrix} |A|I_3,$$

Hence $A(adj\ A) = (adj\ A)A = |A|I_3, A(adj\ A) = (adj\ A)A = |A|I_3,$

15. Find Adjoint of A= $\begin{bmatrix} \mathbf{2} & \mathbf{1} & \mathbf{1} \\ \mathbf{1} & \mathbf{1} & \mathbf{1} \\ \mathbf{1} & \mathbf{-1} & \mathbf{2} \end{bmatrix}$ **and B=** $\begin{bmatrix} \mathbf{1} & \mathbf{2} & \mathbf{3} \\ \mathbf{1} & \mathbf{1} & \mathbf{1} \\ \mathbf{2} & \mathbf{3} & \mathbf{4} \end{bmatrix}$

Solution

$$A = \begin{bmatrix} 2 & 1 & 1 \\ 1 & 1 & 1 \\ 1 & -1 & 2 \end{bmatrix}$$

Cofactor of A

$a_{11} = 1(2+1) = 3 \qquad a_{21} = -1(2+1) = -3 \quad a_{31} = 0$

$a_{21} = -1(2-1) = -1 \quad a_{22} = 1(4-1) = 3 \qquad a_{32} = -1(2-1) = -1$

$a_{31} = 1(-1-1) = -2 \quad a_{23} = -1(-2-1) = 3 \quad a_{33} = 1(2-1) = 1$

Cofactor of A= $\begin{bmatrix} 3 & -1 & -2 \\ -3 & 3 & 3 \\ 0 & -1 & 1 \end{bmatrix}$ Adjoint of A= $\begin{bmatrix} 3 & -3 & 0 \\ -1 & 3 & -1 \\ -2 & 3 & 1 \end{bmatrix}$

$$B = \begin{bmatrix} 1 & 2 & 3 \\ 1 & 1 & 1 \\ 2 & 3 & 4 \end{bmatrix}$$

Cofactor of B-

$b_{11} = 1(4-3) = 1 \qquad b_{21} = -1(8-9) = 1 \quad b_{31} = 1(2-3) = -1$

$b_{12} = -1(4-2) = -2 \quad b_{22} = 1(4-6) = -2 \quad b_{32} = -1(1-3) = 2$

$b_{13} = 1(3-2) = 1 \qquad b_{23} = -1(3-4) = 1 \quad b_{33} = 1(1-2) = -1$

Cofactor of B= $\begin{bmatrix} 1 & -2 & 1 \\ 1 & -2 & 1 \\ -1 & 2 & -1 \end{bmatrix}$ Adjoint of B= $\begin{bmatrix} 1 & 1 & -1 \\ -2 & -2 & 2 \\ 1 & 1 & -1 \end{bmatrix}$

Inverse of a Square Matrix

Let A be a square matrix of order n. If there exists a square matrix B of order n such that AB = BA = In, then the matrix B is called an inverse of A.

If a square matrix has an inverse, then it is unique.

Let A be square matrix of order n. Then, A−1 exists if and only if A is non-singular.

16. If A = $\begin{bmatrix} a & b \\ c & d \end{bmatrix}$ **is non-singular, find** A^{-1}

Solution

We first find adj A. By definition, we get

$$adj\ A = \begin{bmatrix} M_{11} & -M_{12} \\ -M_{21} & M_{22} \end{bmatrix}^T = \begin{bmatrix} d & -c \\ -b & a \end{bmatrix}^T = \begin{bmatrix} d & -b \\ -c & a \end{bmatrix}$$

Since A is non-singular, $|A| = ad - bc \neq 0$

As $A^{-1} = \frac{1}{|A|} adj\ A$, we get $A^{-1} = \frac{1}{ab-bc} \begin{bmatrix} d & -b \\ -c & a \end{bmatrix}$

17. Find the inverse of the matrix $\begin{bmatrix} 2 & -1 & 3 \\ -5 & 3 & 1 \\ -3 & 2 & 3 \end{bmatrix}$

Solution

Let $A = \begin{bmatrix} 2 & -1 & 3 \\ -5 & 3 & 1 \\ -3 & 2 & 3 \end{bmatrix}$ $Then\ |A| = \begin{vmatrix} 2 & -1 & 3 \\ -5 & 3 & 1 \\ -3 & 2 & 3 \end{vmatrix}$

$$= 2(7) + (-12) + 3(-1) = -1 \neq 0$$

Therefore, $A^{-1}A^{-1}$ exists. Now, we get

$$adj\ A = \begin{bmatrix} +\begin{vmatrix} 3 & 1 \\ 2 & 3 \end{vmatrix} & -\begin{vmatrix} -5 & 1 \\ -3 & 3 \end{vmatrix} & +\begin{vmatrix} -5 & 3 \\ -3 & 1 \end{vmatrix} \\ -\begin{vmatrix} -1 & 3 \\ 2 & 3 \end{vmatrix} & +\begin{vmatrix} 2 & 3 \\ -3 & 3 \end{vmatrix} & -\begin{vmatrix} 2 & -1 \\ -3 & 2 \end{vmatrix} \\ +\begin{vmatrix} -1 & 3 \\ 3 & 1 \end{vmatrix} & -\begin{vmatrix} 2 & 3 \\ -5 & 1 \end{vmatrix} & +\begin{vmatrix} 2 & -1 \\ -5 & 3 \end{vmatrix} \end{bmatrix}^T$$

$$= \begin{bmatrix} 7 & 12 & -1 \\ 9 & 15 & -1 \\ -10 & -17 & 1 \end{bmatrix}^T$$

$$= \begin{bmatrix} 7 & 9 & -10 \\ 12 & 15 & -17 \\ -1 & -1 & 1 \end{bmatrix}$$

Hence, $A^{-1} = \frac{1}{|A|}(adj\ A)A^{-1} = \frac{1}{|A|}(adj\ A)$

$$= \frac{1}{-1}\begin{bmatrix} 7 & 9 & -10 \\ 12 & 15 & -17 \\ -1 & -1 & 1 \end{bmatrix}$$

$$= \begin{bmatrix} -7 & -9 & 10 \\ -12 & -15 & 17 \\ 1 & 1 & -1 \end{bmatrix}$$

18. If adj $A = \begin{bmatrix} -1 & 2 & 2 \\ 1 & 1 & 2 \\ 2 & 2 & 1 \end{bmatrix}$ find A^{-1}.

Solution

We compute $|adj\ A| = \begin{vmatrix} -1 & 2 & 2 \\ 1 & 1 & 2 \\ 2 & 2 & 1 \end{vmatrix} = 9$

So, we get $A^{-1} = \pm \frac{1}{\sqrt{|adjA|}} adjA$

$$= \pm\frac{1}{\sqrt{9}}\begin{vmatrix} -1 & 2 & 2 \\ 1 & 1 & 2 \\ 2 & 2 & 1 \end{vmatrix} = \pm\frac{1}{3}\begin{vmatrix} -1 & 2 & 2 \\ 1 & 1 & 2 \\ 2 & 2 & 1 \end{vmatrix}$$

19. Find a matrix *A* if adj (*A*) = $\begin{bmatrix} \mathbf{7} & \mathbf{7} & \mathbf{-7} \\ \mathbf{-1} & \mathbf{11} & \mathbf{7} \\ \mathbf{11} & \mathbf{5} & \mathbf{7} \end{bmatrix}$

Solution

First, we find $|adj\ (A)| = \begin{vmatrix} 7 & 7 & -7 \\ -1 & 11 & 7 \\ 11 & 5 & 7 \end{vmatrix}$

$$= 7(77 - 35) - 7(-7 - 77) - 7(-5 - 121) = 1764 > 0$$

So, we get

$$A = \pm\frac{1}{\sqrt{|adj\ A|}}adj(adj\ A)$$

$$= \pm\frac{1}{\sqrt{1764}}\begin{bmatrix} +(77-35) & -7(-7-77) & +(-5-121) \\ -(49+35) & +(49+77) & -(35-77) \\ +(49+77) & -(49-7) & +(77+7) \end{bmatrix}^T$$

$$= \pm\frac{1}{42}\begin{bmatrix} 42 & 84 & -126 \\ -84 & 126 & 42 \\ 126 & -42 & 84 \end{bmatrix}^T$$

$$= \pm\begin{bmatrix} 1 & -2 & 3 \\ 2 & 3 & -1 \\ -3 & 1 & 2 \end{bmatrix}$$

20. Verify the property $(A^T)^{-1} = (A^{-1})^T$ *with* A = $\begin{bmatrix} \mathbf{2} & \mathbf{9} \\ \mathbf{1} & \mathbf{7} \end{bmatrix}$.

Solution

For the given A, we get |A |= (2) (7) - (9)(1) = 14 – 9 = 5

So, $A^{-1} = \frac{1}{5}\begin{bmatrix} 7 & -9 \\ -1 & 2 \end{bmatrix} = \begin{bmatrix} \frac{7}{5} & -\frac{9}{5} \\ -\frac{1}{5} & \frac{2}{5} \end{bmatrix}$

Then, $(A^{-1})^T = \begin{bmatrix} \frac{7}{5} & -\frac{1}{5} \\ -\frac{9}{5} & \frac{2}{5} \end{bmatrix} = \frac{1}{5}\begin{bmatrix} 7 & -1 \\ -9 & 2 \end{bmatrix}$

From (1) and (2), we get $(A^{-1}) = (A^T)^{-1}$. Thus, we have verified the given property.

For the given A, we get $A^T = \begin{bmatrix} 2 & 1 \\ 9 & 7 \end{bmatrix}$. So $|A^T| = (2)(7) - (1)(9) = 5$

Then, $(A^T)^{-1} = \frac{1}{5}\begin{bmatrix} 7 & -1 \\ -9 & 2 \end{bmatrix}$

21. Verify $(AB)^{-1} = B^{-1} A^{-1}$ with A= $\begin{bmatrix} 0 & -3 \\ 1 & 4 \end{bmatrix}$ and B= $\begin{bmatrix} -2 & -3 \\ 0 & -1 \end{bmatrix}$

Solution

We get

$$AB = \begin{bmatrix} 0 & -3 \\ 1 & 4 \end{bmatrix}\begin{bmatrix} -2 & -3 \\ 0 & -1 \end{bmatrix}$$

$$= \begin{bmatrix} 0+0 & 0+3 \\ -2+0 & -3-4 \end{bmatrix}$$

$$= \begin{bmatrix} 0 & 3 \\ -2 & -7 \end{bmatrix}$$

$$(AB)^{-1} = \frac{1}{0+6}\begin{bmatrix} -7 & -3 \\ 2 & 0 \end{bmatrix}$$

$$= \frac{1}{6}\begin{bmatrix} -7 & -3 \\ 2 & 0 \end{bmatrix}$$

$$A^{-1} = \frac{1}{0+3}\begin{bmatrix} 4 & 3 \\ -1 & 0 \end{bmatrix}$$

$$= \frac{1}{3}\begin{bmatrix} 4 & 3 \\ -1 & 0 \end{bmatrix}$$

$$B^{-1} = \frac{1}{2-0}\begin{bmatrix} -1 & 3 \\ 0 & -2 \end{bmatrix}$$

$$= \frac{1}{2}\begin{bmatrix} -1 & 3 \\ 0 & -2 \end{bmatrix}$$

$$B^{-1}A^{-1} = \frac{1}{2}\begin{bmatrix} -1 & 3 \\ 0 & -2 \end{bmatrix}\frac{1}{3}\begin{bmatrix} 4 & 3 \\ -1 & 0 \end{bmatrix} = \frac{1}{6}\begin{bmatrix} -7 & -3 \\ 2 & 0 \end{bmatrix}$$

As the matrices in (1) and (2) are same, $(AB)^{-1} = B^{-1} A^{-1}$ is verified.

22. If $A + B = \begin{bmatrix} 1 & 2 \\ 5 & 6 \end{bmatrix}$ and $A - B = \begin{bmatrix} -3 & 4 \\ -1 & -2 \end{bmatrix}$ then AB=?

Solution

If $A + B = \begin{bmatrix} 1 & 2 \\ 5 & 6 \end{bmatrix} \ldots\ldots.(1)$

$A - B = \begin{bmatrix} -3 & 4 \\ -1 & -2 \end{bmatrix} \ldots\ldots(2)$

$formula\ 1 + formula\ 2$

$(A + B) + (A - B) = 2A$

$\therefore 2A = \begin{bmatrix} 1 & 2 \\ 5 & -6 \end{bmatrix} + \begin{bmatrix} -3 & 4 \\ -1 & -2 \end{bmatrix}$

$2A = \begin{bmatrix} -2 & 6 \\ 4 & -8 \end{bmatrix}$

$A = \begin{bmatrix} -1 & 3 \\ 2 & -4 \end{bmatrix}$

Now,

$A + B = \begin{bmatrix} 1 & 2 \\ 5 & -6 \end{bmatrix}$

$B = \begin{bmatrix} 1 & 2 \\ 5 & -6 \end{bmatrix} - \begin{bmatrix} -1 & 3 \\ 2 & -4 \end{bmatrix}$

$B = \begin{bmatrix} 2 & -1 \\ 3 & -2 \end{bmatrix}$

$A * B = \begin{bmatrix} -1 & 3 \\ 2 & -4 \end{bmatrix} * \begin{bmatrix} 2 & -1 \\ 3 & -2 \end{bmatrix}$

$= \begin{bmatrix} -2 + 9 & +1 - 6 \\ 4 - 12 & -2 + 8 \end{bmatrix}$

$= \begin{bmatrix} 7 & -5 \\ -8 & 6 \end{bmatrix}$

23. If $A = \begin{bmatrix} 3 & 3 & 4 \\ 2 & -3 & 4 \\ 0 & -1 & 1 \end{bmatrix}$ **Prove that** $A^3 = A^{-1}$

Solution

$$A = \begin{bmatrix} 3 & 3 & 4 \\ 2 & -3 & 4 \\ 0 & -1 & 1 \end{bmatrix}$$

$$|A| = 3(-3+4) - 3(2-0) + 4(-2-0)$$

$$|A| = 3(1) - 3(2) - 8 = -11$$

$$A^3 = A^2 * A$$

$$A^2 = \begin{bmatrix} 3 & 3 & 4 \\ 2 & -3 & 4 \\ 0 & -1 & 1 \end{bmatrix} * \begin{bmatrix} 3 & 3 & 4 \\ 2 & -3 & 4 \\ 0 & -1 & 1 \end{bmatrix}$$

$$A^2 = \begin{bmatrix} 9+6+0 & 9-9-4 & 12+12+4 \\ 6-6+0 & 6+9-4 & 8-18+4 \\ 0-2+0 & 0+3-1 & 0-4+1 \end{bmatrix}$$

$$A^2 = \begin{bmatrix} 15 & -4 & 28 \\ 0 & 11 & 6 \\ -2 & 2 & -3 \end{bmatrix}$$

$$A^3 = \begin{bmatrix} 15 & -4 & 28 \\ 0 & 11 & -6 \\ -2 & 2 & -3 \end{bmatrix} * \begin{bmatrix} 3 & 3 & 4 \\ 2 & -3 & 4 \\ 0 & -1 & 1 \end{bmatrix}$$

$$A^3 = \begin{bmatrix} 45-8+0 & 45+12-28 & 60-16+28 \\ 0+22+0 & 0-33+6 & 0+44-6 \\ -6+4+0 & -6-6+3 & -8+8-3 \end{bmatrix}$$

$$A^3 = \begin{bmatrix} 37 & 29 & 72 \\ 22 & -27 & 38 \\ -2 & -9 & -3 \end{bmatrix}$$

Now,

$$A^{-1} = \frac{Adj\,[A]}{|A|}$$

$$\therefore Cofactor\ matrix = \begin{bmatrix} 1 & -2 & -2 \\ 1 & 3 & 3 \\ 24 & -4 & -15 \end{bmatrix}$$

$$\therefore Adjoint\ matrix = \begin{bmatrix} 1 & 1 & 24 \\ -2 & 3 & -4 \\ -2 & 3 & -15 \end{bmatrix}$$

$$|A| = 11$$

$$A^{-1} = \frac{\begin{bmatrix} 1 & 1 & 24 \\ -2 & 3 & -4 \\ -2 & 3 & -15 \end{bmatrix}}{11}$$

24. If $A = \begin{bmatrix} 4 & 3 \\ 2 & 5 \end{bmatrix}$, find x and y such that $A^2 + xA + yI_2 = O_{2,}$ Hence, find A^{-1}.

Solution

Since

$$A^2 = \begin{bmatrix} 4 & 3 \\ 2 & 5 \end{bmatrix}\begin{bmatrix} 4 & 3 \\ 2 & 5 \end{bmatrix} = \begin{bmatrix} 22 & 27 \\ 18 & 31 \end{bmatrix} A^2 = \begin{bmatrix} 4 & 3 \\ 2 & 5 \end{bmatrix}\begin{bmatrix} 4 & 3 \\ 2 & 5 \end{bmatrix} = \begin{bmatrix} 22 & 27 \\ 18 & 31 \end{bmatrix}$$

$$A^2 + xA + yI_2 = O_2 \rightarrow \begin{bmatrix} 22 & 27 \\ 18 & 31 \end{bmatrix} + x\begin{bmatrix} 4 & 3 \\ 2 & 5 \end{bmatrix} + y\begin{bmatrix} 1 & 0 \\ 0 & 1 \end{bmatrix} = \begin{bmatrix} 0 & 0 \\ 0 & 0 \end{bmatrix}$$

$$= \begin{bmatrix} 22 + 4x + y & 27 + 3x \\ 18 + 2x & 31 + 5x + y \end{bmatrix} = \begin{bmatrix} 0 & 0 \\ 0 & 0 \end{bmatrix}$$

So, we get 22 + 4x + y =0, 31+5x+y=0, 27+3x=0 and 18+2x=0

Hence x = −9 and y =14.Then, we get $A^2 - 9A + 14I_2 = O_2$

Post-multiplying this equation by A^{-1}, we get $A - 9I_2 + 14A^{-1} = O_2$. Hence, we get

$$A^{-1} = \frac{1}{14}(9I_2 - A) = \frac{1}{14}\left(9\begin{bmatrix} 1 & 0 \\ 0 & 1 \end{bmatrix} - \begin{bmatrix} 4 & 3 \\ 2 & 5 \end{bmatrix}\right) = \frac{1}{14}\begin{bmatrix} 5 & -3 \\ -2 & 4 \end{bmatrix}$$

25. Find the inverse of the following matrix $\begin{bmatrix} 1 & 2 & 1 \\ 2 & -1 & 2 \\ 1 & 1 & -2 \end{bmatrix}$

Solution

Cofactor of A-

$a_{11} = 1(2-2) = 0$ $\quad a_{21} = -1(-4-1) = 5$ $\quad a_{31} = 1(4+1) = 5$

$a_{12} = -1(-4-2) = 6$ $\quad a_{22} = 1(-2-1) = -3$ $\quad a_{31} = 0a_{31} = 0$

$a_{13} = 1(2+1) = 3$ $\quad a_{23} = -1(1-2) = 1$ $\quad a_{33} = 1(-1-4) = -5$

Adjoint of A= $\begin{bmatrix} 0 & 5 & 5 \\ 6 & -3 & 0 \\ 3 & 1 & -5 \end{bmatrix}$

Formula: $\boldsymbol{A^{-1}} = \dfrac{\boldsymbol{Adj\ A}}{|\boldsymbol{A}|}$

$|A| = 1(2-2) - 2(-4-2) + 1(2+1) = 15$

$$A^{-1} = \frac{1}{15}\begin{bmatrix} 0 & 5 & 5 \\ 6 & -3 & 0 \\ 3 & 1 & -5 \end{bmatrix} = \begin{bmatrix} 0 & 1/3 & 1/3 \\ 2/5 & -1/5 & 0 \\ 2/5 & 1/15 & -1/3 \end{bmatrix}$$

26. Find determinants of following: $\begin{bmatrix} \mathbf{3} & \mathbf{4} & \mathbf{-1} \\ \mathbf{0} & \mathbf{-1} & \mathbf{2} \\ \mathbf{5} & \mathbf{-2} & \mathbf{6} \end{bmatrix}$ **and** $\begin{bmatrix} \mathbf{1} & \mathbf{1} & \mathbf{-1} \\ \mathbf{2} & \mathbf{1} & \mathbf{-2} \\ \mathbf{1} & \mathbf{-1} & \mathbf{1} \end{bmatrix}$

Solution

A= $\begin{bmatrix} 3 & 4 & -1 \\ 0 & -1 & 2 \\ 5 & -2 & 6 \end{bmatrix}$ $|A| = 3(-6+4) - 4(0-10) + (-1)(0+5) = 29$

B= $\begin{bmatrix} 1 & 1 & -1 \\ 2 & 1 & -2 \\ 1 & -1 & 1 \end{bmatrix}$ $|B| = 1(1-2) - 1(2+2) + (-1)(-2-1) = -2$

27. Solve the equation by Matrix method

(a) $\boldsymbol{x + y = 3}$

$\boldsymbol{2x + 3y = 8}$

(b) $\boldsymbol{2x - y = 7}$

$\boldsymbol{3x - 2y = 11}$

Solution

(a) Matrix form= $\begin{bmatrix} 1 & 1 \\ 2 & 3 \end{bmatrix}\begin{bmatrix} x \\ y \end{bmatrix} = \begin{bmatrix} 3 \\ 8 \end{bmatrix}$

$\boldsymbol{Formula: X = A^{-1}B}$

A= $\begin{bmatrix} 1 & 1 \\ 2 & 3 \end{bmatrix}$;

Adj A= $\begin{bmatrix} 3 & -1 \\ -2 & 1 \end{bmatrix}$; $|A| = 3 - 2 = 1;$ $\quad A^{-1} = \begin{bmatrix} 3 & -1 \\ -2 & 1 \end{bmatrix}$

$$X = A^{-1}B = \begin{bmatrix} 3 & -1 \\ -2 & 1 \end{bmatrix}\begin{bmatrix} 3 \\ 8 \end{bmatrix} = \begin{bmatrix} 1 \\ 2 \end{bmatrix}$$

Therefore, X=1 and Y=2

Matrix form= $\begin{bmatrix} 2 & -1 \\ 3 & -2 \end{bmatrix}\begin{bmatrix} x \\ y \end{bmatrix} = \begin{bmatrix} 7 \\ 11 \end{bmatrix}$

$\boldsymbol{Formula: X = A^{-1}B}$

Adj B= $\begin{bmatrix} -2 & 1 \\ -3 & 2 \end{bmatrix}$; $|A| = -4 + 3 = -1; A^{-1} = \begin{bmatrix} 2 & -1 \\ 3 & -2 \end{bmatrix}$

$$X = A^{-1}B = \begin{bmatrix} 2 & -1 \\ 3 & -2 \end{bmatrix}\begin{bmatrix} 7 \\ 11 \end{bmatrix} = \begin{bmatrix} 3 \\ -1 \end{bmatrix}$$

28. Find the rank of following matrices

i. $\mathbf{A} = \begin{bmatrix} \mathbf{1} & \mathbf{2} & \mathbf{3} \\ \mathbf{2} & \mathbf{3} & \mathbf{1} \\ \mathbf{-2} & \mathbf{-3} & \mathbf{-1} \end{bmatrix}$

Solution

$$|A| = 1(-3+3) - 2(-2+2) + 3(-6+6) = 0$$

Determinant of a minor in matrix A= $\begin{vmatrix} 1 & 2 \\ 2 & 3 \end{vmatrix} = 3 - 4 = -1$

Since the minor does not come zero the rank will be the size of minor i.e., 2 (rank of A)

ii. $B = \begin{bmatrix} 1 & 2 & 3 \\ 2 & 3 & 1 \\ 3 & 1 & 2 \end{bmatrix}$

Solution

$$|B| = 1(6-1) - 2(4-3) + 3(2-9) = -18$$

$$|B| \neq 0|B| \neq 0$$

As the determinant of the matrix B is not equal to zero, the rank of the matrix will be size of the matrix i.e., 3

iii. $C = \begin{bmatrix} 3 & -1 & 2 \\ -6 & 2 & 4 \\ -3 & 1 & 2 \end{bmatrix}$

Solution

$$|C| = 3(4-4) + 1(-12+12) + 2(-6+6)$$

$$|C| = 0|C| = 0$$

Determinant of a minor in matrix C= $\begin{vmatrix} -1 & 2 \\ 1 & 2 \end{vmatrix} = -2 - 2 = -4 \neq 0$

Since the minor does not come zero the rank will be the size of minor i.e., 2 (rank of C)

iv. $D = \begin{bmatrix} 0 & 1 & 2 \\ -1 & 0 & 3 \\ -2 & -3 & 0 \end{bmatrix}$

Solution

$$[D] = 0 - 1(6) + 2(3) = 0$$

Determinant of a minor in matrix D= $\begin{vmatrix} 0 & 3 \\ -3 & 0 \end{vmatrix} = 0 + 9 = 9 \neq 0$

Since the minor does not come zero the rank will be the size of minor i.e., 2 (rank of D)

v. $E = \begin{bmatrix} 6 & 1 & 3 & 8 \\ 16 & 4 & 12 & 15 \\ 10 & 3 & 9 & 7 \end{bmatrix}$

Solution

$$R_1/6 \rightarrow \begin{bmatrix} 1 & 1/6 & 1/2 & 4/3 \\ 16 & 4 & 12 & 15 \\ 10 & 3 & 9 & 7 \end{bmatrix}$$

$$R_2 - 16R_1 \& R_3 - 10R_1 \rightarrow \begin{bmatrix} 1 & 1/6 & 1/2 & 4/3 \\ 0 & 4/3 & 4 & -19/3 \\ 0 & 4/3 & 4 & -19/3 \end{bmatrix}$$

$$R_3 - R_2 \rightarrow \begin{bmatrix} 1 & 1/6 & 1/2 & 4/3 \\ 0 & 4/3 & 4 & -19/3 \\ 0 & 0 & 0 & 0 \end{bmatrix}$$

There are two non-zero rows, hence rank of E is 2

29. Find the rank of given matrix

$$A = \begin{bmatrix} 4 & 2 & 6 & -1 \\ 6 & 1 & 3 & 8 \\ 16 & 4 & 12 & 15 \\ 10 & 3 & 9 & 7 \end{bmatrix}$$

$$\sim \begin{bmatrix} 4 & 2 & 6 & -1 \\ 6 & 1 & 3 & 8 \\ 16 & 4 & 12 & 15 \\ 10 & 3 & 9 & 4 \end{bmatrix}$$

$$C1 \rightarrow \frac{C1}{2}, C3 \rightarrow \frac{C3}{3}$$

$$\begin{bmatrix} 2 & 2 & 2 & -1 \\ 3 & 1 & 1 & 8 \\ 8 & 4 & 4 & 15 \\ 5 & 3 & 3 & 7 \end{bmatrix}$$

$C2 \leftrightarrow C1$

$$\sim \begin{bmatrix} 1 & 2 & 2 & 2 \\ 8 & 1 & 1 & 3 \\ 15 & 4 & 4 & 8 \\ 7 & 3 & 3 & 5 \end{bmatrix}$$

$R1 \to R2 + 821$

$R3 \to R3 + 15R1$

$R4 \to R4 + 7R1$

$$\sim \begin{bmatrix} -1 & 2 & 2 & 2 \\ 0 & 17 & 17 & 19 \\ 0 & 34 & 34 & 38 \\ 0 & 17 & 17 & 19 \end{bmatrix}$$

$R3 \to R3/2$

$$\sim \begin{bmatrix} -1 & 2 & 2 & 2 \\ 0 & 17 & 17 & 19 \\ 0 & 17 & 17 & 19 \\ 0 & 17 & 17 & 19 \end{bmatrix}$$

$R2 \to R2 - R1$

$R3 \to R3 - R2$

$R4 \to R4 - R3$

$$\begin{bmatrix} -1 & 2 & 2 & 2 \\ 0 & 0 & 0 & 0 \\ 0 & 0 & 0 & 0 \\ 0 & 0 & 0 & 0 \end{bmatrix}$$

Therefore, rank of the matrix is 1.

The rank of the matrix

$$\boldsymbol{A} = \begin{bmatrix} \mathbf{2} & \mathbf{3} & \mathbf{1} & \mathbf{4} \\ \mathbf{0} & \mathbf{1} & \mathbf{2} & \mathbf{-1} \\ \mathbf{0} & \mathbf{-2} & \mathbf{-4} & \mathbf{2} \end{bmatrix}$$

Solution

$$A = \begin{bmatrix} 2 & 3 & 1 & 4 \\ 0 & 1 & 2 & -1 \\ 0 & -2 & -4 & 2 \end{bmatrix}$$

$R3 \rightarrow R3 + 2R2$

$$A = \begin{bmatrix} 2 & 3 & 1 & 4 \\ 0 & 1 & 2 & -1 \\ 0 & 0 & 0 & 0 \end{bmatrix}$$

$therefore, rank\ of\ the\ given\ matrix\ is\ 2$

30. Find λ if rank of A = $\begin{bmatrix} 2 & -1 & 3 \\ 4 & 7 & \lambda \\ 1 & 4 & 5 \end{bmatrix} = 2$

Solution

Rank of matrix is 2 means the determinant of A is zero

$2(35 - 4\lambda) + 1(20 - \lambda) + 3(16 - 7) = 0$

$70 - 8\lambda + 20 - \lambda + 48 - 21 = 0$

$9\lambda = 117$

$\lambda = 13$

APPLICATIONS OF MATRICES: SOLVING SYSTEM OF LINEAR EQUATIONS

A system of linear equations having at least one solution is said to be consistent. A system of linear equations having no solution is said to be inconsistent.

Remark

If the number of the equations of a system of linear equations is equal to the number of unknowns of the system, then the coefficient matrix A of the system is a square matrix. Further, if A is a non-singular matrix, then the solution of system of equations can be found by any one of the following methods: (i) matrix inversion method, (ii) Cramer's rule, (iii) Gaussian elimination method.

Matrix Inversion Method

This method can be applied only when the coefficient matrix is a square matrix and non-singular.

31. Solve the following system of linear equations, using matrix inversion method: $5x + 2y = 3, 3x + 2y = 5$.

Solution

The matrix form of the system is AX = B, where

$$A = \begin{bmatrix} 5 & 2 \\ 3 & 2 \end{bmatrix}, X = \begin{bmatrix} X \\ Y \end{bmatrix}, B = \begin{bmatrix} 3 \\ 5 \end{bmatrix}$$

We find $|A| = \begin{vmatrix} 5 & 2 \\ 3 & 2 \end{vmatrix} = 10 - 6 = 4 \neq 0$. So, A^{-1} exists and $A^{-1} = \frac{1}{4}\begin{bmatrix} 2 & -2 \\ -3 & 5 \end{bmatrix}$

Then, applying the formula $X = A^{-1}B$ we get

$$\begin{bmatrix} x \\ y \end{bmatrix} = \frac{1}{4}\begin{bmatrix} 2 & -2 \\ -3 & 5 \end{bmatrix}\begin{bmatrix} 3 \\ 5 \end{bmatrix}$$

$$= \frac{1}{4}\begin{bmatrix} 6 - 10 \\ -9 + 25 \end{bmatrix} = \frac{1}{4}\begin{bmatrix} -4 \\ 16 \end{bmatrix} = \begin{bmatrix} -1 \\ 4 \end{bmatrix}$$

So, the solution is (x= -1, y=4)

Cramer's Rule

This rule can be applied only when the coefficient matrix is a square matrix and non-singular.

$$\boldsymbol{x_1 = \frac{\Delta_1}{\Delta}, x_2 = \frac{\Delta_2}{\Delta}, x_3 = \frac{\Delta_3}{\Delta}}$$

32. Solve, by Cramer's rule, the system of equations

$x_1 - x_2 = 3, 2x_1 + 3x_2 + 4x_3 = 17, x_2 + 2x_3 = 7$.

Solution

First, we evaluate the determinants

$$\Delta = \begin{vmatrix} 1 & -1 & 0 \\ 2 & 3 & 4 \\ 0 & 1 & 0 \end{vmatrix} = 6 \neq 0,$$

$$\Delta_1 = \begin{vmatrix} 3 & -1 & 0 \\ 17 & 3 & 4 \\ 7 & 1 & 2 \end{vmatrix} = 12,$$

$$\Delta_2 = \begin{vmatrix} 1 & 3 & 0 \\ 2 & 17 & 4 \\ 0 & 7 & 2 \end{vmatrix} = -6$$

$$\Delta_3 = \begin{vmatrix} 1 & -1 & 3 \\ 2 & 3 & 17 \\ 0 & 1 & 7 \end{vmatrix} = 24$$

By Cramer's rule, we get

$$x_1 = \frac{\Delta_1}{\Delta} = \frac{12}{6} = 2,$$

$$x_2 = \frac{\Delta_2}{\Delta} = -\frac{6}{6} = -1$$

$$x_3 = \frac{24}{6} = 4$$

So, the solution is (x1 = 2, x2 = - 1, x3 = 4).

33. The system of equations

$$\boldsymbol{x + 4y - 2z = 3}$$

$$\boldsymbol{3x + y + 5z = 7}$$

$$\boldsymbol{2x + 3y + z = 5}$$

Has unique solution, no solution or many solutions?

Solution

$$A = \begin{bmatrix} 1 & 4 & -2 \\ 3 & 1 & 5 \\ 2 & 3 & 1 \end{bmatrix}, x = \begin{bmatrix} x \\ y \\ z \end{bmatrix}, B = \begin{bmatrix} 3 \\ 7 \\ 5 \end{bmatrix}$$

$$[AB] = \begin{bmatrix} 1 & 4 & -2 & 3 \\ 3 & 1 & 5 & 7 \\ 2 & 3 & 1 & 5 \end{bmatrix}$$

$R2 \to R2 - 3R1$

$$[AB] = \begin{bmatrix} 1 & 4 & -2 & 3 \\ 0 & -11 & 11 & -2 \\ 2 & 3 & 1 & 5 \end{bmatrix}$$

$R3 \to R3 - 2R1$

$$[AB] = \begin{bmatrix} 1 & 4 & -2 & 3 \\ 0 & -11 & 11 & -2 \\ 0 & -5 & 5 & -1 \end{bmatrix}$$

$C2 \rightarrow C2 + C3$

$$[AB] = \begin{bmatrix} 1 & 2 & -2 & 3 \\ 0 & 0 & 11 & -2 \\ 0 & 0 & 5 & -1 \end{bmatrix}$$

$R3 \rightarrow 11R3 - 5R2$

$$[AB] = \begin{bmatrix} 1 & 2 & -2 & 3 \\ 0 & 0 & 11 & -2 \\ 0 & 0 & 0 & -1 \end{bmatrix}$$

3RD row is non-zero row, so the rank of matrix is 3

$\therefore \rho(AB)=3$

Now, rank of A

$$\rho(A) = \begin{bmatrix} 1 & 4 & -2 \\ 3 & 1 & 5 \\ 2 & 3 & 1 \end{bmatrix}$$

$$|A| = 1(1-15) - 4(3-10) - 2(9-2)$$

$$|A| = -14 + 28 - 14 = 0$$

$$\rightarrow \begin{vmatrix} 1 & 5 \\ 3 & 1 \end{vmatrix}$$

$$= (1 - 15) = -14$$

$\therefore$ rank of matrix A $=\rho(A) = 2$

Now,

$\rho(AB) \neq \rho(A)$

So, this system of equation has no solution.

34. For what value of λ the matrix has non-zero solution-

$$\mathbf{3x - y + z = 0}$$

$$\mathbf{15x - 6y + 5z = 015x - 6y + 5z = 0}$$

$$\mathbf{\lambda x - 2y + 2z = 0\lambda x - 2y + 2z = 0}$$

Solution

For a non-zero solution determinant of matrix is zero

$$\begin{vmatrix} 3 & -1 & 1 \\ 15 & -6 & 5 \\ \lambda & -2 & 2 \end{vmatrix} = 0$$

$$3(-12+10)+1(30-5\lambda)+1(-30+6\lambda)=0$$

$$\lambda = 6\lambda = 6$$

GAUSSIAN ELIMINATION METHOD

This method can be applied even if the coefficient matrix is singular matrix and rectangular matrix. It is essentially the method of substitution which we have already seen. In this method, we transform the augmented matrix of the system of linear equations into row-echelon form and then by back-substitution, we get the solution.

35. Solve the following system of linear equations, by Gaussian elimination method: $4x + 3y + 6z = 25$, $x + 5y + 7z = 13$, $2x + 9y + z = 1$.

Solution

Transforming the augmented matrix to echelon form, we get

$$\begin{bmatrix} 4 & 3 & 6 & \vdots\ 25 \\ 1 & 5 & 7 & \vdots\ 13 \\ 2 & 9 & 1 & \vdots\ 1 \end{bmatrix}$$

$$R_1 \leftrightarrow R_2$$

$$\begin{bmatrix} 1 & 5 & 7 & \vdots\ 13 \\ 1 & 3 & 6 & \vdots\ 25 \\ 2 & 9 & 1 & \vdots\ 1 \end{bmatrix}$$

$$R_2 \to R_2 - 4R_1$$

$$R_2 \to R_3 - 2R_1$$

$$\begin{bmatrix} 1 & 5 & 7 & \vdots\ +13 \\ 0 & -17 & -22 & \vdots\ -27 \\ 0 & -1 & -13 & \vdots\ -25 \end{bmatrix}$$

$$R_2 \to R_2 * (-1)$$

$$R_3 \to R_3 * (-1)$$

$$\begin{bmatrix} 1 & 5 & 7 & \vdots\ 13 \\ 0 & 17 & 22 & \vdots\ 27 \\ 0 & 1 & 13 & \vdots\ 25 \end{bmatrix}$$

$R_3 \rightarrow 17R_3 - R_2$

$$\begin{bmatrix} 1 & 5 & 7 & \vdots 13 \\ 0 & 17 & 22 & \vdots 27 \\ 0 & 0 & 199 & \vdots 398 \end{bmatrix}$$

The equivalent system is written by using the echelon form:

x + 5y + 7z = 13, ... (1)

17y + 22z = 27, ... (2)

199z = 398, ... (3)

FROM (3), we get z=398/199=2

Substituting z = 2, y = -1 in (1), we get x = 13 - 5 × (−1) − 7 × 2 = 4.

So, the solution is (x =4, y = - 1, z = 2).

Note. The above method of going from the last equation to the first equation is called the method of back substitution.

36. The system of equation x+4y-2z = 3 has unique/non zero/infinite solution?

3x+y+5z = 7

2x+3y+z = 5

Solution

$$A = \begin{bmatrix} 1 & 4 & -2 \\ 3 & 1 & 5 \\ 2 & 3 & 1 \end{bmatrix}$$

$|A| = 0$

Determinant of minor $\begin{vmatrix} 1 & 4 \\ 3 & 1 \end{vmatrix} = -11$

SO, rank of A is 2

Let A: B= $\begin{vmatrix} 1 & 4 & -2 & :3 \\ 3 & 1 & 5 & :7 \\ 2 & 3 & 1 & :5 \end{vmatrix}$

$$R_3 \rightarrow R_3 - 2R_1 \& R_2 \rightarrow R_2 - 2R_1 = \begin{bmatrix} 1 & 4 & -2: 3 \\ 0 & -11 & 11: -2 \\ 0 & -5 & 5: 1 \end{bmatrix}$$

$$R_3 \to \frac{R_3}{5} \,\&R_2 \to \frac{R_2}{11} = \begin{bmatrix} 1 & 4 & -2:3 \\ 0 & -1 & 1:-\frac{2}{11} \\ 0 & -1 & 1:\frac{1}{5} \end{bmatrix}$$

$$R_3 \to R_3 - R_1 = \begin{bmatrix} 1 & 4 & -2 & :3 \\ 0 & -1 & 1 & :-\frac{2}{11} \\ 0 & 0 & 0 & :-\frac{1}{53} \end{bmatrix}$$

Rank of Augmented matrix is 3

Rank of A is not equal to rank of A: B

Hence there will be no solution

37. Solve the equation - x+y+z=9

2x+5y+2z=52

2x+y-z=0

Solution

$$\Delta = \begin{bmatrix} 1 & 1 & 1 \\ 2 & 5 & 2 \\ 2 & 1 & -1 \end{bmatrix} = -9$$

$$\Delta x = \begin{bmatrix} 9 & 1 & 1 \\ 52 & 5 & 2 \\ 0 & 1 & -1 \end{bmatrix} = 41$$

$$\Delta y = \begin{bmatrix} 1 & 9 & 1 \\ 2 & 52 & 2 \\ 2 & 0 & -1 \end{bmatrix} = -102$$

$$\Delta z = \begin{bmatrix} 1 & 1 & 9 \\ 2 & 5 & 52 \\ 2 & 1 & 0 \end{bmatrix} = -20$$

$$x = \frac{\Delta x}{\Delta} = \frac{-41}{9}$$

$$y = \frac{\Delta y}{\Delta} = \frac{102}{9}$$

$$z = \frac{\Delta z}{\Delta} = \frac{20}{9}$$

38. Check $x+2y+3z=1$; $2x+y+3z=2$ & $4x+5y+9z=4$ has infinite solution.

Solution

$$A=\begin{bmatrix}1 & 2 & 3\\ 2 & 1 & 3\\ 4 & 5 & 9\end{bmatrix}$$

$|A|=0$

Determinant of minor $\begin{bmatrix}1 & 2\\ 2 & 1\end{bmatrix}=-3$

Rank of A is 2

$$A{:}\,B=\begin{bmatrix}1 & 2 & 3 & :1\\ 2 & 1 & 3 & :2\\ 4 & 5 & 9 & :4\end{bmatrix}$$

$$R_2\to R_2-2R_1 \& R_3\to R_3-4R_1=\begin{bmatrix}1 & 2 & 3 & :1\\ 0 & -3 & -3 & :0\\ 0 & -3 & -3 & :0\end{bmatrix}$$

$$R_3\to R_3-R_2=\begin{bmatrix}1 & 2 & 3 & :1\\ 0 & -3 & -3 & :0\\ 0 & 0 & 0 & :0\end{bmatrix}$$

Rank of augmented matrix A: B is 2

Here rank of A and rank of A: B are same. So, there is infinite solution

39. For what value of λ the system of equation has consistent solution

$$X+y+z=6;\quad x+2y+3z=10\quad \&\quad x+2y+\lambda z=12$$

Solution

For consistent solution |A|≠0

$|A|=2\lambda-6-\lambda+3\neq 0$

$\lambda-3\neq 0$

$\lambda\neq 3$

For all values except 3 for λ will give consistent solution for above system of equation

40. For what value of λ system is trivial solution

$$X - 2y + z = 0;\ \ 2x - y + 3z = 0\ \ \&\ \ \lambda x + y - 2z = 0$$

Solution

For trivial solution IAI ≠0

$$\begin{vmatrix} 1 & -2 & 1 \\ 2 & -1 & 3 \\ \lambda & 1 & -2 \end{vmatrix} \neq 0$$

$$= -1 + 2(-4 - 3\lambda) + \lambda + 2 \neq 0$$

$$= \lambda \neq \frac{7}{5}$$

For all values except 7/5 for λ will give trival solution for above system of equation

41. If A+B = $\begin{bmatrix} 1 & 2 \\ 5 & -6 \end{bmatrix}$ **and A-B =** $\begin{bmatrix} -3 & 4 \\ -1 & -2 \end{bmatrix}$ **then AB =?**

Solution

$$\begin{bmatrix} a11 + b11 & a12 + b12 \\ a21 + b21 & a22 + b22 \end{bmatrix} = \begin{bmatrix} 1 & 2 \\ 5 & -6 \end{bmatrix}$$

$$\begin{bmatrix} a11 - b11 & a12 - b12 \\ a21 - b21 & a22 - b22 \end{bmatrix} = \begin{bmatrix} -3 & 4 \\ -1 & -2 \end{bmatrix}$$

Add both matrices, $2\begin{bmatrix} a11 & a12 \\ a21 & a11 \end{bmatrix} = \begin{bmatrix} -2 & 6 \\ 4 & -8 \end{bmatrix}$

$$A = \begin{bmatrix} -1 & 3 \\ 2 & -4 \end{bmatrix} \&\ B = \begin{bmatrix} 2 & -1 \\ 3 & -2 \end{bmatrix}$$

$$AB = \begin{bmatrix} 7 & -5 \\ -8 & 6 \end{bmatrix}$$

42. Find inverse of $\begin{bmatrix} \alpha & \gamma \\ \beta & \delta \end{bmatrix}\begin{bmatrix} \alpha & \gamma \\ \beta & \delta \end{bmatrix}$

Solution

$$A^{-1} = \frac{Adj\ A}{|A|} = \frac{1}{\alpha\delta - \beta\gamma}\begin{bmatrix} \delta & -\gamma \\ -\beta & \alpha \end{bmatrix}$$

43. If $A = \begin{bmatrix} 2x & 0 \\ x & x \end{bmatrix}$ and $A^{-1} = \begin{bmatrix} 1 & 0 \\ -1 & 2 \end{bmatrix}$ find X

Solution

$$A^{-1} = \frac{Adj\ A}{|A|}$$

Here, $Adj\ A = \begin{bmatrix} x & 0 \\ -x & 2x \end{bmatrix}$ $and\ |A| = 2x^2$

$$A^{-1} = \frac{1}{2x^2}\begin{bmatrix} x & 0 \\ -x & 2x \end{bmatrix} = \begin{bmatrix} 1 & 0 \\ -1 & 2 \end{bmatrix}$$

$$\begin{bmatrix} \frac{1}{2x} & 0 \\ \frac{-1}{2x} & \frac{1}{x} \end{bmatrix} = \begin{bmatrix} 1 & 0 \\ -1 & 2 \end{bmatrix}$$

$$x = \frac{1}{2}$$

44. If $A = \begin{bmatrix} 1 & 2 \\ 4 & 3 \end{bmatrix}$ and

$f(a) = 2x^2 - 4x + 5$, *prove that* $f(a) = \begin{bmatrix} 19 & 8 \\ 16 & 27 \end{bmatrix}$

Solution

$$f(a) = 2\begin{bmatrix} 1 & 2 \\ 4 & 3 \end{bmatrix}^2 - 4\begin{bmatrix} 1 & 2 \\ 4 & 3 \end{bmatrix} + 5I$$

$$= 2\begin{bmatrix} 9 & 8 \\ 16 & 17 \end{bmatrix} - \begin{bmatrix} 4 & 8 \\ 16 & 12 \end{bmatrix} + \begin{bmatrix} 5 & 0 \\ 0 & 5 \end{bmatrix}$$

$$= \begin{bmatrix} 18 & 16 \\ 32 & 34 \end{bmatrix} - \begin{bmatrix} 4 & 8 \\ 16 & 12 \end{bmatrix} + \begin{bmatrix} 5 & 0 \\ 0 & 5 \end{bmatrix}$$

$$= \begin{bmatrix} 19 & 8 \\ 16 & 27 \end{bmatrix}$$

45. The system of equation has unique/non zero / infinite solution

$$x + 2y + 3z = 1 \; ; \; 2x + y + 3z = 2 \quad \& \quad 4x + 5y + 9z = 4$$

Solution

$$A = \begin{bmatrix} 1 & 2 & 3 \\ 2 & 1 & 3 \\ 4 & 5 & 9 \end{bmatrix}; \; |A| = 0; \; Rank\ of\ minor = 2$$

Rank of A = 2

Augmented matrix $A{:}\,B = \begin{bmatrix} 1 & 2 & 3{:}\,1 \\ 2 & 1 & 3{:}\,2 \\ 4 & 5 & 9{:}\,4 \end{bmatrix} A{:}\,B = \begin{bmatrix} 1 & 2 & 3{:}\,1 \\ 2 & 1 & 3{:}\,2 \\ 4 & 5 & 9{:}\,4 \end{bmatrix}$

$$R_3 \to R_3 - 4R_1 \& R_2 \to R_2 - 2R_1 = \begin{bmatrix} 1 & 2 & 3{:}\,1 \\ 0 & -3 & -3{:}\,0 \\ 0 & -3 & -3{:}\,0 \end{bmatrix}$$

$$R_3 \to R_3 - R_2 = \begin{bmatrix} 1 & 4 & 2{:}\,3 \\ 0 & -3 & -3.0 \\ 0 & 0 & 0{:}\,0 \end{bmatrix} \; i, e., rank\ of\ A{:}\,B = 2$$

Here rank of A: B is equal with rank of A which is 2 and it is less tank unknown factors (x,y,z) which is 3. So, there will be infinite solution.

46. The system of equation has unique/non zero / infinite solution?

$$x + 2y - z = 3 \; ; \; 2x - 2y + 3z = 2 \quad \& \quad 3x - y + 2z = 1$$

Solution

$$A = \begin{bmatrix} 1 & 2 & -1 \\ 2 & -2 & 3 \\ 3 & -1 & 2 \end{bmatrix}; \; |A| = 15; So\ rank\ of\ A\ is\ 3$$

Augmented matrix $A{:}\,B = \begin{bmatrix} 1 & 2 & -1{:}\,3 \\ 2 & -2 & 3{:}\,2 \\ 3 & -1 & 2{:}\,1 \end{bmatrix}$

$$R_3 \to R_3 - 3R_1 \& R_2 \to R_2 - 2R_1 = \begin{bmatrix} 1 & 2 & -1{:}\,3 \\ 0 & -6 & 5{:}\,-4 \\ 0 & -7 & 5{:}\,-8 \end{bmatrix}$$

$$R_3 \to R_3 - {}^{7}/_{6} R_2 = \begin{bmatrix} 1 & 2 & -1:3 \\ 0 & -6 & 5:-4 \\ 0 & 0 & -\frac{5}{6}:-\frac{10}{3} \end{bmatrix}$$

Rank of A: B is 3

Here, rank of A and A: B are equal and both ranks are equal with unknown numbers 3 (x,y,z). So, there will be a unique solution.

47. For what value of λ, A has trivial solution

$$\boldsymbol{x - 2y + 2z = 0; \quad 2x - 1y + 3z = 0} \quad \& \quad \lambda x + y - 2z = 0$$

Solution

For trivial solution, rank of A should be less than 3, i.e., IAI = 0

$$A = \begin{bmatrix} 1 & -2 & 2 \\ 2 & -1 & 3 \\ \lambda & 1 & -2 \end{bmatrix}$$

$$|A| = 1(2-3) + 2(-4-3\lambda) + 2(2+\lambda) = 0$$

$$-1 - 8 - 6\lambda + 4 + 2\lambda = 0$$

$$\lambda = {}^{-5}/_{4}$$

48. For what value of λ system is inconsistent-

$$\boldsymbol{x + y + z = 6; \quad x + 2y + 3z = 10} \quad \& \quad x + 2y + \lambda z = \boldsymbol{12}$$

Solution

$$A = \begin{bmatrix} 1 & 1 & 1 \\ 1 & 2 & 3 \\ 1 & 2 & \lambda \end{bmatrix}$$

For inconsistent solution the rank of A should be less than number of unknown numbers *i,e.,* 2. For rank 2 the determinant of A should be zero

$$|A| = \begin{vmatrix} 1 & 1 & 1 \\ 1 & 2 & 3 \\ 1 & 2 & \lambda \end{vmatrix} = 0$$

$$1(2\lambda - 6) - 1(\lambda - 3) + 1(2-2) = 0$$

$$\lambda = 3$$

49. Find the eigen values and eigen vectors of following matrices.

(a) $\begin{bmatrix} 1 & 2 \\ 0 & 3 \end{bmatrix}$

$\boldsymbol{Eigen\ value\ formula} = |A - \lambda I| = 0$

$A - \begin{bmatrix} \lambda & 0 \\ 0 & \lambda \end{bmatrix} = 0$

$= \begin{bmatrix} 1 & 2 \\ 0 & 3 \end{bmatrix} - \begin{bmatrix} \lambda & 0 \\ 0 & \lambda \end{bmatrix} = 0$

$= \begin{vmatrix} 1-\lambda & 2 \\ 0 & 3-\lambda \end{vmatrix} = 0$

$= (1-\lambda)(3-\lambda) = 0$

$= 3 - \lambda - 3\lambda + \lambda^2 = 0$

$= \lambda^2 - 4\lambda + 3 = 0$

$= (\lambda - 3)(\lambda - 1) = 0$

$\lambda = 3\ \&\ 1$

So, the eigen values are 3 and 1

50. For eigen vectors we have to follow $Av = \lambda v$

By simplifying we will get, $(A - \lambda I)v = 0$

For case 1 (λ=3)

$(A - \lambda I)v = 0$

$= \left[\begin{bmatrix} 1 & 2 \\ 0 & 3 \end{bmatrix} - \begin{bmatrix} \lambda & 0 \\ 0 & \lambda \end{bmatrix}\right]\begin{bmatrix} v_1 \\ v_2 \end{bmatrix} = 0$

$= \begin{bmatrix} 1-3 & 2 \\ 0 & 3-3 \end{bmatrix}\begin{bmatrix} v_1 \\ v_2 \end{bmatrix} = 0$

$= \begin{bmatrix} -2 & 2 \\ 0 & 0 \end{bmatrix}\begin{bmatrix} v_1 \\ v_2 \end{bmatrix} = 0$

$= -2v_1 + 2v_2 = 0$

Let $v_2 = t,$

Then for equation $-2v_1 + 2v_2 = 0$

$= -2v_1 + 2t = 0$

$= v = t$

Therefore, $v = \begin{bmatrix} 1 \\ 1 \end{bmatrix} t$

For case 2 (λ=1)

$(\boldsymbol{A} - \boldsymbol{\lambda I})\boldsymbol{v} = \boldsymbol{0}$

$= \left[\begin{bmatrix} 1 & 2 \\ 0 & 3 \end{bmatrix} - \begin{bmatrix} \lambda & 0 \\ 0 & \lambda \end{bmatrix}\right]\begin{bmatrix} v_1 \\ v_2 \end{bmatrix} = 0$

$= \begin{bmatrix} 1-1 & 2 \\ 0 & 3-1 \end{bmatrix}\begin{bmatrix} v_1 \\ v_2 \end{bmatrix} = 0$

$= \begin{bmatrix} 0 & 2 \\ 0 & 2 \end{bmatrix}\begin{bmatrix} v_1 \\ v_2 \end{bmatrix} = 0$

$= 2v_2 + 2v_2 = 0$

Let $v_1 = t$,

Then we found that $v_2 = 0$

Therefore, $v = \begin{bmatrix} 1 \\ 0 \end{bmatrix} t$

Eigen Values	*Eigen Vectors*
3	1,1
1	1,0

(b) $\begin{bmatrix} \mathbf{1} & \mathbf{2} \\ \mathbf{4} & \mathbf{3} \end{bmatrix}$

Solution

$A - \begin{bmatrix} \lambda & 0 \\ 0 & \lambda \end{bmatrix} = 0$

$= \begin{bmatrix} 1 & 2 \\ 4 & 3 \end{bmatrix} - \begin{bmatrix} \lambda & 0 \\ 0 & \lambda \end{bmatrix} = 0$

$$= \begin{vmatrix} 1-\lambda & 2 \\ 4 & 3-\lambda \end{vmatrix} = 0$$

$$= (1-\lambda)(3-\lambda) - 8 = 0$$

$$= 3 - \lambda - 3\lambda + \lambda^2 - 8 = 0$$

$$= \lambda^2 - 4\lambda - 5 = 0$$

$$= (\lambda - 5)(\lambda + 1) = 0$$

$$\lambda = 5 \,\&\, -1$$

So, the eigen values are 5 and -1

For eigen vectors we have to follow $\boldsymbol{Av = \lambda v}$

By simplifying we will get, $\boldsymbol{(A - \lambda I)v = 0}$

For case 1 (λ=5)

$$\boldsymbol{(A - \lambda I)v = 0}$$

$$= \left[\begin{bmatrix} 1 & 2 \\ 4 & 3 \end{bmatrix} - \begin{bmatrix} \lambda & 0 \\ 0 & \lambda \end{bmatrix}\right]\begin{bmatrix} v_1 \\ v_2 \end{bmatrix} = 0$$

$$= \begin{bmatrix} 1-5 & 2 \\ 4 & 3-5 \end{bmatrix}\begin{bmatrix} v_1 \\ v_2 \end{bmatrix} = 0$$

$$= \begin{bmatrix} -4 & 2 \\ 4 & -2 \end{bmatrix}\begin{bmatrix} v_1 \\ v_2 \end{bmatrix} = 0$$

$$= -4v_1 + 2v_2 = 0 \,\& 4v_1 - 2v_2 = 0$$

Let $v_2 = t$,

Then for equation $4v_1 - 2v_2 = 0$

$$v_1 = \frac{t}{2}$$

Therefore, $v = \begin{bmatrix} 1/2 \\ 1 \end{bmatrix} t$

For case 2 (λ=-1)

$$\boldsymbol{(A - \lambda I)v = 0}$$

$$= \left[\begin{bmatrix}1 & 2\\4 & 3\end{bmatrix} - \begin{bmatrix}\lambda & 0\\0 & \lambda\end{bmatrix}\right]\begin{bmatrix}v_1\\v_2\end{bmatrix} = 0$$

$$= \begin{bmatrix}1-(-1) & 2\\4 & 3-(-1)\end{bmatrix}\begin{bmatrix}v_1\\v_2\end{bmatrix} = 0$$

$$= \begin{bmatrix}2 & 2\\4 & 4\end{bmatrix}\begin{bmatrix}v_1\\v_2\end{bmatrix} = 0$$

$$= 2v_1 + 2v_2 = 0 \quad \&4v_1 + 4v_2 = 0$$

Let $v_2 = t,$

Then for equation $4v_1 + 4v_2 = 0$

$v_1 = -t$

Therefore, $v = \begin{bmatrix}-1\\1\end{bmatrix}t$

Eigen Values	*Eigen Vectors*
5	1/2,1
-1	-1,0

(c) $\begin{bmatrix}\mathbf{5} & \mathbf{2}\\\mathbf{2} & \mathbf{5}\end{bmatrix}\begin{bmatrix}\mathbf{5} & \mathbf{2}\\\mathbf{2} & \mathbf{5}\end{bmatrix}$

Solution

$$A - \begin{bmatrix}\lambda & 0\\0 & \lambda\end{bmatrix} = 0$$

$$= \begin{bmatrix}5 & 2\\2 & 5\end{bmatrix} - \begin{bmatrix}\lambda & 0\\0 & \lambda\end{bmatrix} = 0$$

$$= \begin{vmatrix}5-\lambda & 2\\2 & 5-\lambda\end{vmatrix} = 0$$

$$= (5-\lambda)(5-\lambda) - 4 = 0$$

$$= 25 - 5\lambda - 5\lambda + \lambda^2 - 8 = 0$$

$$= \lambda^2 - 10\lambda + 21 = 0$$

$= (\lambda - 7)(\lambda - 3) = 0$

$\lambda = 7 \ \& \ 3$

For case 1 (λ=7)

$(\boldsymbol{A} - \boldsymbol{\lambda I})\boldsymbol{v} = \boldsymbol{0}$

$$= \left[\begin{bmatrix} 5 & 2 \\ 2 & 5 \end{bmatrix} - \begin{bmatrix} \lambda & 0 \\ 0 & \lambda \end{bmatrix}\right]\begin{bmatrix} v_1 \\ v_2 \end{bmatrix} = 0$$

$$= \begin{bmatrix} 5-7 & 2 \\ 2 & 5-7 \end{bmatrix}\begin{bmatrix} v_1 \\ v_2 \end{bmatrix} = 0$$

$$= \begin{bmatrix} -2 & 2 \\ 2 & -2 \end{bmatrix}\begin{bmatrix} v_1 \\ v_2 \end{bmatrix} = 0$$

$= -2v_1 + 2v_2 = 0 \quad \& 2v_1 - 2v_2 = 0$

Let $v_2 = t,$

Then for equation $2v_1 - 2v_2 = 0$

$v_1 = t$

Therefore, $v = \begin{bmatrix} 1 \\ 1 \end{bmatrix} t$

For case 2 (λ=3)

$(\boldsymbol{A} - \boldsymbol{\lambda I})\boldsymbol{v} = \boldsymbol{0}$

$$= \left[\begin{bmatrix} 5 & 2 \\ 2 & 5 \end{bmatrix} - \begin{bmatrix} \lambda & 0 \\ 0 & \lambda \end{bmatrix}\right]\begin{bmatrix} v_1 \\ v_2 \end{bmatrix} = 0$$

$$= \begin{bmatrix} 5-3 & 2 \\ 2 & 5-3 \end{bmatrix}\begin{bmatrix} v_1 \\ v_2 \end{bmatrix} = 0$$

$$= \begin{bmatrix} 2 & 2 \\ 2 & 2 \end{bmatrix}\begin{bmatrix} v_1 \\ v_2 \end{bmatrix} = 0$$

$= 2v_1 + 2v_2 = 0 \quad \& 2v_1 + 2v_2 = 0$

Let $v_2 = t,$

Then for equation $2v_1 + 2v_2 = 0$

$v_1 = -t$

Therefore, $v = \begin{bmatrix} -1 \\ 1 \end{bmatrix} t$

Eigen Values	*Eigen Vectors*
7	1,1
3	-1,1

(d) $\begin{bmatrix} \mathbf{2} & \mathbf{4} \\ \mathbf{0} & \mathbf{3} \end{bmatrix} \begin{bmatrix} \mathbf{2} & \mathbf{4} \\ \mathbf{0} & \mathbf{3} \end{bmatrix}$

Solution

$$A - \begin{bmatrix} \lambda & 0 \\ 0 & \lambda \end{bmatrix} = 0$$

$$= \begin{bmatrix} 2 & 4 \\ 0 & 3 \end{bmatrix} - \begin{bmatrix} \lambda & 0 \\ 0 & \lambda \end{bmatrix} = 0$$

$$= \begin{vmatrix} 2-\lambda & 4 \\ 0 & 3-\lambda \end{vmatrix} = 0$$

$$= (2-\lambda)(3-\lambda) - 0 = 0$$

$$= 6 - 2\lambda - 3\lambda + \lambda^2 = 0$$

$$= \lambda^2 - 5\lambda + 6 = 0$$

$$= (\lambda - 2)(\lambda - 3) = 0$$

$$\lambda = 2 \ \& \ 3$$

For case 1 (λ=2)

$$(\boldsymbol{A} - \boldsymbol{\lambda I})\boldsymbol{v} = \mathbf{0}$$

$$= \left[\begin{bmatrix} 2 & 4 \\ 0 & 2 \end{bmatrix} - \begin{bmatrix} \lambda & 0 \\ 0 & \lambda \end{bmatrix}\right] \begin{bmatrix} v_1 \\ v_2 \end{bmatrix} = 0$$

$$= \begin{bmatrix} 2-2 & 4 \\ 0 & 3-2 \end{bmatrix} \begin{bmatrix} v_1 \\ v_2 \end{bmatrix} = 0$$

$$= \begin{bmatrix} 0 & 4 \\ 0 & 1 \end{bmatrix} \begin{bmatrix} v_1 \\ v_2 \end{bmatrix} = 0$$

$= 4v_2 = 0 \quad \& v_2 = 0$

Let $v_1 = t,$

Then for equation $v_2 = 0$

Therefore, $v = \begin{bmatrix} 1 \\ 0 \end{bmatrix} t$

For case 2 (λ=3)

$(\boldsymbol{A} - \boldsymbol{\lambda I})\boldsymbol{v} = \boldsymbol{0}$

$= \left[\begin{bmatrix} 2 & 4 \\ 0 & 3 \end{bmatrix} - \begin{bmatrix} \lambda & 0 \\ 0 & \lambda \end{bmatrix}\right] \begin{bmatrix} v_1 \\ v_2 \end{bmatrix} = 0$

$= \begin{bmatrix} 2-3 & 4 \\ 0 & 3-3 \end{bmatrix} \begin{bmatrix} v_1 \\ v_2 \end{bmatrix} = 0$

$= \begin{bmatrix} -1 & 4 \\ 0 & 0 \end{bmatrix} \begin{bmatrix} v_1 \\ v_2 \end{bmatrix} = 0$

$= -v_1 + 4v_2 = 0$

Let $v_2 = t,$

Then for equation $-v_1 + 4v_2 = 0$

$v_1 = 4t$

Therefore, $v = \begin{bmatrix} 4 \\ 1 \end{bmatrix} t$

Eigen Values	*Eigen Vectors*
2	1,0
3	4,1

(e) $\begin{bmatrix} -2 & -4 & 2 \\ -2 & 1 & 2 \\ 4 & 2 & 5 \end{bmatrix}$

Solution

$|A - \lambda I| = 0$

$$\begin{bmatrix} -2-\lambda & -4 & 2 \\ -2 & 1-\lambda & 2 \\ 4 & 2 & 5-\lambda \end{bmatrix} = 0$$

$$(-2-\lambda)[(1-\lambda)(5-\lambda)-4]+4[(-2)(5-\lambda)-8]+2(-4-4+4\lambda)=0$$

$$(-2-\lambda)(5-\lambda-5\lambda+\lambda^2-4)+4(-10+2\lambda-8)-8-8+8\lambda=0$$

$$-10+2\lambda+10\lambda-2\lambda^2+8-5\lambda+\lambda^2+5\lambda^2-\lambda^3+4\lambda-16+8\lambda=0$$

$$-90+27\lambda+4\lambda^2-\lambda^3=0$$

$$\lambda^3-4\lambda^2-27\lambda+90=0$$

Factorization of above equation will give $\lambda_1 = 3 \quad \lambda_2 = 6 \quad \lambda_3 = -5$

Case 1 (λ=3)

$$(\boldsymbol{A}-\boldsymbol{\lambda I})\boldsymbol{v} = \boldsymbol{0}$$

$$\begin{bmatrix} -2-3 & -4 & 2 \\ -2 & 1-3 & 2 \\ 4 & 2 & 5-3 \end{bmatrix}\begin{bmatrix} v_1 \\ v_2 \\ v_3 \end{bmatrix} = 0$$

$$\begin{bmatrix} -5 & -4 & 2 \\ -2 & -2 & 2 \\ 4 & 2 & 2 \end{bmatrix}\begin{bmatrix} v_1 \\ v_2 \\ v_3 \end{bmatrix} = 0$$

$$A = \begin{bmatrix} -5 & -4 & 2 \\ -2 & -2 & 2 \\ 4 & 2 & 2 \end{bmatrix} \quad R_2 - {}^2/_5\, R_1 \& R_3 + {}^4/_5\, R_1 \rightarrow \begin{bmatrix} -5 & -4 & 2 \\ 0 & -{}^2/_5 & {}^6/_5 \\ 4 & -{}^6/_5 & {}^{18}/_5 \end{bmatrix}$$

$$R_3 - 3R_2 \rightarrow \begin{bmatrix} -5 & -4 & 2 \\ 0 & -{}^2/_5 & {}^6/_5 \\ 0 & 0 & 0 \end{bmatrix}$$

$$\begin{bmatrix} -5 & -4 & 2 \\ 0 & -{}^2/_5 & {}^6/_5 \\ 0 & 0 & 0 \end{bmatrix}\begin{bmatrix} v_1 \\ v_2 \\ v_3 \end{bmatrix} = 0$$

$$-5v_1 - 4v_2 + 2v_3 = 0 \qquad \& \qquad -{}^2/_5\, v_2 + {}^6/_5\, v_3 = 0$$

Let $v_3 = t,$

Then for equation $-{}^{2}/_{5}\, v_2 + {}^{6}/_{5}\, v_3 = 0$

$v_2 = 3t$

Put values of v_2 &v_3 in $-5v_1 - 4v_2 + 2v_3 = 0$

$-5v_1 = 12t - 2t$

$v_1 = -2t$

Therefore, $v = \begin{bmatrix} -2 \\ 3 \\ 1 \end{bmatrix} t$

Case 2 (λ=6)

$$(\boldsymbol{A} - \boldsymbol{\lambda I})\boldsymbol{v} = \boldsymbol{0}$$

$$\begin{bmatrix} -2-6 & -4 & 2 \\ -2 & 1-6 & 2 \\ 4 & 2 & 5-6 \end{bmatrix} \begin{bmatrix} v_1 \\ v_2 \\ v_3 \end{bmatrix} = 0$$

$$\begin{bmatrix} -8 & -4 & 2 \\ -2 & -5 & 2 \\ 4 & 2 & -1 \end{bmatrix} \begin{bmatrix} v_1 \\ v_2 \\ v_3 \end{bmatrix} = 0$$

$$A = \begin{bmatrix} -8 & -4 & 2 \\ -2 & -5 & 2 \\ 4 & 2 & -1 \end{bmatrix} \qquad 4R_2 - R_1 \& \ 2R_3 + R_1 \rightarrow \begin{bmatrix} -8 & -4 & 2 \\ 0 & -16 & 6 \\ 0 & 0 & 0 \end{bmatrix}$$

$$\begin{bmatrix} -8 & -4 & 2 \\ 0 & -16 & 6 \\ 0 & 0 & 0 \end{bmatrix} \begin{bmatrix} v_1 \\ v_2 \\ v_3 \end{bmatrix} = 0$$

$-8v_1 - 4v_2 + 2v_3 = 0$ & $-16v_2 + 6v_3 = 0$

Let $v_3 = t$,

Then for equation $-16v_2 + 6v_3 = 0$

$v_2 = {}^{3t}/_{8}$

Put values of v_2 &v_3 in $-8v_1 - 4v_2 + 2v_3 = 0$

$-8v_1 = -{}^{t}/_{2} -8v_1 = -{}^{t}/_{2}$

$$v_1 = {}^{t}/_{16}$$

Therefore, $v = \begin{bmatrix} {}^{1}/_{16} \\ {}^{3}/_{8} \\ 1 \end{bmatrix} t$

Case 3 (λ=-5)

$$(\boldsymbol{A} - \boldsymbol{\lambda I})\boldsymbol{v} = \boldsymbol{0}$$

$$\begin{bmatrix} -2+5 & -4 & 2 \\ -2 & 1+5 & 2 \\ 4 & 2 & 5+5 \end{bmatrix} \begin{bmatrix} v_1 \\ v_2 \\ v_3 \end{bmatrix} = 0$$

$$\begin{bmatrix} 3 & -4 & 2 \\ -2 & 6 & 2 \\ 4 & 2 & 10 \end{bmatrix} \begin{bmatrix} v_1 \\ v_2 \\ v_3 \end{bmatrix} = 0$$

$$A = \begin{bmatrix} -8 & -4 & 2 \\ -2 & -5 & 2 \\ 4 & 2 & -1 \end{bmatrix} \quad R_2 + {}^{2}/_{3}\, R_1 \& R_3 - {}^{4}/_{3}\, R_1 \rightarrow \begin{bmatrix} 3 & -4 & 2 \\ 0 & {}^{10}/_{3} & {}^{10}/_{3} \\ 0 & {}^{22}/_{3} & {}^{22}/_{3} \end{bmatrix}$$

$$R_3 - {}^{11}/_{5}\, R_3 \rightarrow \begin{bmatrix} 3 & -4 & 2 \\ 0 & {}^{10}/_{3} & {}^{10}/_{3} \\ 0 & 0 & 0 \end{bmatrix}$$

$$3v_1 - 4v_2 + 2v_3 = 0 \& {}^{10}/_{3}\, v_2 + {}^{10}/_{3}\, v_3 = 0$$

Let $v_3 = t,$

Then for equation ${}^{10}/_{3}\, v_2 + {}^{10}/_{3}\, v_3 = 0$

$v_2 = -t$

Put values of $v_2 \& v_3$ in $3v_1 - 4v_2 + 2v_3 = 0$

$3v_1 = -2t - 4t$

$v_1 = -2t$

Therefore, $v = \begin{bmatrix} -2 \\ -1 \\ 1 \end{bmatrix} tv = \begin{bmatrix} -2 \\ -1 \\ 1 \end{bmatrix} t$

Eigen Values	*Eigen Vectors*
3	-2, 3 & 1
6	1/16, 3/8, 1
-5	-2, -1 & 1

(f) $\begin{bmatrix} \mathbf{1} & \mathbf{-2} \\ \mathbf{3} & \mathbf{-4} \end{bmatrix}\begin{bmatrix} \mathbf{1} & \mathbf{-2} \\ \mathbf{3} & \mathbf{-4} \end{bmatrix}$

Solution

$$A - \begin{bmatrix} \lambda & 0 \\ 0 & \lambda \end{bmatrix} = 0$$

$$= \begin{bmatrix} 1 & -2 \\ 3 & -4 \end{bmatrix} - \begin{bmatrix} \lambda & 0 \\ 0 & \lambda \end{bmatrix} = 0$$

$$= \begin{vmatrix} 1-\lambda & -2 \\ 3 & -4-\lambda \end{vmatrix} = 0$$

$$= (1-\lambda)(-4-\lambda) + 6 = 0$$

$$= -4 - \lambda + 4\lambda + \lambda^2 + 6 = 0$$

$$= \lambda^2 + 3\lambda + 2 = 0$$

$$= (\lambda + 1)(\lambda + 2) = 0$$

$$\lambda = -1 \,\&- 2$$

For case 1 (λ=-1)

$$(\boldsymbol{A} - \boldsymbol{\lambda I})\boldsymbol{v} = \mathbf{0}$$

$$= \left[\begin{bmatrix} 1 & -2 \\ 3 & -4 \end{bmatrix} - \begin{bmatrix} \lambda & 0 \\ 0 & \lambda \end{bmatrix}\right]\begin{bmatrix} v_1 \\ v_2 \end{bmatrix} = 0$$

$$= \begin{bmatrix} 1+1 & -2 \\ 3 & -4+1 \end{bmatrix}\begin{bmatrix} v_1 \\ v_2 \end{bmatrix} = 0$$

$$= \begin{bmatrix} 2 & -2 \\ 3 & -3 \end{bmatrix} \begin{bmatrix} v_1 \\ v_2 \end{bmatrix} = 0$$

$2v_1 - 2v_2 = 0\&\quad 3v_1 - 3v_2 = 0$

Let $v_2 = t,$

Then for equation $2v_1 - 2v_2 = 0$

$v_1 = t$

Therefore, $v = \begin{bmatrix} 1 \\ 1 \end{bmatrix} t$

For case 2 (λ=-2)

$(\boldsymbol{A} - \boldsymbol{\lambda I})\boldsymbol{v} = \boldsymbol{0}$

$$= \left[\begin{bmatrix} 1 & -2 \\ 3 & -4 \end{bmatrix} - \begin{bmatrix} \lambda & 0 \\ 0 & \lambda \end{bmatrix} \right] \begin{bmatrix} v_1 \\ v_2 \end{bmatrix} = 0$$

$$= \begin{bmatrix} 1+2 & -2 \\ 3 & -4+2 \end{bmatrix} \begin{bmatrix} v_1 \\ v_2 \end{bmatrix} = 0$$

$$= \begin{bmatrix} 3 & -2 \\ 3 & -2 \end{bmatrix} \begin{bmatrix} v_1 \\ v_2 \end{bmatrix} = 0$$

$= 3v_1 - 2v_2 = 0$

Let $v_2 = t,$

Then for equation $3v_1 - 2v_2 = 0$

$v_1 = {}^{2}/_{3}\, t$

Therefore, $v = \begin{bmatrix} 2/3 \\ 1 \end{bmatrix} t$

Eigen Values	*Eigen Vectors*
-1	1 & 1
-2	2/3 & 1

51. If A & B are three rowed square matrices such that A=3B and |A|= λ |B|. Then find the value of λ.

Solution

As we know when we take common for a matrix then it comes from all element of a matrix.

Let, A= $\begin{bmatrix} 3 & 0 & 0 \\ 0 & 3 & 0 \\ 0 & 0 & 3 \end{bmatrix}$ then if we take common from A, then A= $3\begin{bmatrix} 1 & 0 & 0 \\ 0 & 1 & 0 \\ 0 & 0 & 1 \end{bmatrix}$

But during the determinant the common element reduce from each row

i.e., A= $\begin{vmatrix} 3 & 0 & 0 \\ 0 & 3 & 0 \\ 0 & 0 & 3 \end{vmatrix} = 3 \times 3 \times 3 \begin{vmatrix} 1 & 0 & 0 \\ 0 & 1 & 0 \\ 0 & 0 & 1 \end{vmatrix} = 27 \begin{vmatrix} 1 & 0 & 0 \\ 0 & 1 & 0 \\ 0 & 0 & 1 \end{vmatrix}$

The common element is 27. So, value of λ is 27

52. If $A = \begin{bmatrix} \cos\theta & -\sin\theta & 0 \\ \sin\theta & \cos\theta & 0 \\ 0 & 0 & 1 \end{bmatrix}$, then prove than $A^{-1} = A^t$

Solution

$$A^{-1} = \frac{Adj\ A}{|A|}$$

Cofactor of A= $\begin{bmatrix} \cos\theta & -\sin\theta & 0 \\ \sin\theta & \cos\theta & 0 \\ 0 & 0 & 1 \end{bmatrix}$ Adj (A)= $\begin{bmatrix} \cos\theta & \sin\theta & 0 \\ -\sin\theta & \cos\theta & 0 \\ 0 & 0 & 1 \end{bmatrix}$

$$|A| = \begin{vmatrix} \cos\theta & -\sin\theta & 0 \\ \sin\theta & \cos\theta & 0 \\ 0 & 0 & 1 \end{vmatrix} = \cos\theta\,(\cos\theta) + \sin\theta\,(\sin\theta) + 0$$

$$= Sin^2\theta + Cos^2\theta = 1$$

$$A^{-1} = \frac{Adj\ A}{|A|} = \frac{Adj\ A}{1} = \begin{bmatrix} \cos\theta & \sin\theta & 0 \\ -\sin\theta & \cos\theta & 0 \\ 0 & 0 & 1 \end{bmatrix}$$

Hence proved $A^{-1} = A^t A^{-1} = A^t$

Elementary Transformation

A non-zero matrix E is said to be in a **row-echelon form** if:

1. All zero rows of E occur below every non-zero row of E.
2. The first non-zero element in any row i of E occurs in the jth column of E, then all other entries in the jth column of E below the first non-zero element of row i are zeros.
3. The first non-zero entry in the ith row of E lies to the left of the first non-zero entry in $(i+1)^{th}$ row of E .

Note

A non-zero matrix is in a row-echelon form if all zero rows occur as bottom rows of the matrix, and if the first non-zero element in any lower row occurs to the right of the first non- zero entry in the higher row.

The following matrices are in row-echelon form:

53. Reduce the matrix $\begin{bmatrix} 3 & -1 & 2 \\ -6 & 2 & 4 \\ -3 & 1 & 2 \end{bmatrix}$ to a row-echelon form.

Solution

$$\begin{bmatrix} 3 & -1 & 2 \\ -6 & 2 & 4 \\ -3 & 1 & 2 \end{bmatrix}$$

$R_2 \to R_2 + 2R_1$

$R_3 \to R_3 + R_1$

$$\begin{bmatrix} 3 & -1 & 2 \\ 0 & 0 & 8 \\ 0 & 0 & 4 \end{bmatrix}$$

$R_3 \to R_3 - \frac{1}{2}R_2$

$$\begin{bmatrix} 3 & -1 & 2 \\ 0 & 0 & 8 \\ 0 & 0 & 0 \end{bmatrix}$$

Note

$R_2 \to R_2/8$

$$\begin{bmatrix} 3 & -1 & 2 \\ 0 & 0 & 1 \\ 0 & 0 & 0 \end{bmatrix}$$

This is also a row-echelon form of the given matrix.

So, a row-echelon form of a matrix is not necessarily unique.

54. Reduce the matrix $\begin{bmatrix} 0 & 3 & 1 & 6 \\ -1 & 0 & 2 & 5 \\ 4 & 2 & 0 & 0 \end{bmatrix}$ to a row-echelon form.

Solution

$$\begin{bmatrix} 0 & 3 & 1 & 6 \\ -1 & 0 & 2 & 5 \\ 4 & 2 & 0 & 0 \end{bmatrix}$$

$R_1 \leftrightarrow R_2$

$$\begin{bmatrix} -1 & 0 & 2 & 5 \\ 0 & 3 & 1 & 6 \\ 4 & 2 & 0 & 0 \end{bmatrix}$$

$R_3 \to R_3 + 4R_1$

$$\begin{bmatrix} -1 & 0 & 2 & 5 \\ 0 & 3 & 1 & 6 \\ 0 & 2 & 8 & 20 \end{bmatrix}$$

$R_3 \to R_3 - \frac{2}{3}R_2$

$$\begin{bmatrix} -1 & 0 & 2 & 5 \\ 0 & 3 & 1 & 6 \\ 0 & 0 & \frac{22}{3} & 16 \end{bmatrix}$$

$R_3 \to 3R_2$

$$\begin{bmatrix} -1 & 0 & 2 & 5 \\ 0 & 3 & 1 & 6 \\ 0 & 0 & 22 & 48 \end{bmatrix}$$

55. Convert given matrices into Echelon form

(a) $\begin{bmatrix} \mathbf{4} & \mathbf{3} & \mathbf{7} \\ \mathbf{1} & \mathbf{1} & \mathbf{5} \\ \mathbf{4} & \mathbf{5} & \mathbf{7} \end{bmatrix}$

Solution

$$R_3 \to R_3 - R_1 \& R_1 \to \frac{R_1}{4} = \begin{bmatrix} 1 & {}^3/_4 & {}^7/_4 \\ 1 & 1 & 5 \\ 0 & 2 & 0 \end{bmatrix}$$

$$R_2 \to R_2 - R_1 \& R_1 \to R_1 - 3R_2 = \begin{bmatrix} 1 & 0 & -8 \\ 1 & {}^1/_4 & {}^{13}/_4 \\ 0 & 2 & 0 \end{bmatrix}$$

$$R_3 \to R_3 - 8R_2 \& R_2 \to 3R_2 = \begin{bmatrix} 1 & 0 & -8 \\ 0 & 1 & 13 \\ 0 & 0 & -26 \end{bmatrix}$$

$$R_3 \to \frac{R_3}{-26} \& R_1 \to R_1 + 8R_3 = \begin{bmatrix} 1 & 0 & 0 \\ 0 & 1 & 13 \\ 0 & 0 & 1 \end{bmatrix}$$

$$R_2 \to R_2 - 13R_3 = \begin{bmatrix} 1 & 0 & 0 \\ 0 & 1 & 0 \\ 0 & 0 & 1 \end{bmatrix}$$

Matrix in Echelon form for A is $\begin{bmatrix} 1 & 0 & 0 \\ 0 & 1 & 0 \\ 0 & 0 & 1 \end{bmatrix}$

(b) $\begin{bmatrix} \mathbf{5} & \mathbf{-6} & \mathbf{-7} & \mathbf{7} \\ \mathbf{6} & \mathbf{-4} & \mathbf{10} & \mathbf{-34} \\ \mathbf{2} & \mathbf{4} & \mathbf{-3} & \mathbf{29} \end{bmatrix}$

Solution

$$R_1 \to \frac{R_1}{5} \& \; R_2 \to R_2 - 6R_1 = \begin{bmatrix} 1 & {}^{-6}/_5 & {}^{-7}/_5 & {}^7/_5 \\ 0 & {}^{16}/_5 & {}^{92}/_5 & -34 \\ 2 & 4 & -3 & 29 \end{bmatrix}$$

$$R_3 \to R_3 - 2R_1 = \begin{bmatrix} 1 & -6/5 & -7/5 & 7/5 \\ 0 & 16/5 & 92/5 & -34 \\ 2 & 32/5 & -1/5 & 131/5 \end{bmatrix}$$

$$R_3 \to R_3 - 2R_1 = \begin{bmatrix} 1 & -6/5 & -7/5 & 7/5 \\ 0 & 16/5 & 92/5 & -34 \\ 0 & 0 & -37 & 111 \end{bmatrix}$$

$$R_2 \to 5R_2/16 = \begin{bmatrix} 1 & -6/5 & -7/5 & 7/5 \\ 1 & 1 & 23/4 & -53/4 \\ 0 & 0 & -37 & 111 \end{bmatrix}$$

$$R_1 \to R_1 + 6R_2/5 = \begin{bmatrix} 1 & 0 & 11/2 & -29/2 \\ 1 & 1 & 23/4 & -53/4 \\ 0 & 0 & -37 & 111 \end{bmatrix}$$

$$R_3 \to R_3/_{-37} \& R_1 \to R_1 - 11R_3/2 = \begin{bmatrix} 1 & 0 & 0 & 2 \\ 1 & 1 & 23/4 & -53/4 \\ 0 & 0 & 1 & -3 \end{bmatrix}$$

$$R_2 \to R_2 - 23R_3/4 = \begin{bmatrix} 1 & 0 & 0 & 2 \\ 0 & 1 & 0 & 4 \\ 0 & 0 & 1 & -3 \end{bmatrix}$$

Matrix in Echelon form for B is $\begin{bmatrix} 1 & 0 & 0 & 2 \\ 0 & 1 & 0 & 4 \\ 0 & 0 & 1 & -3 \end{bmatrix}$

56. Solve the equation by converting to Echelon form

$$x + 4y - z = 4$$

$$2x + 5y + 8z = 15$$

$$x + 3y - 3z = 1$$

Solution

Augmented matrix (A|B) = $\begin{bmatrix} 1 & 4 & -1 & |4 \\ 2 & 5 & 8 & |15 \\ 1 & 3 & -3 & |1 \end{bmatrix}$

$$R_3 \to 2R_3 - R_2 \& R_2 \to R_2 - 2R_1 = \begin{bmatrix} 1 & 4 & -1 & |4 \\ 0 & -3 & 10 & |7 \\ 0 & 1 & -14 & |-13 \end{bmatrix}$$

$$R_3 \to R_3 \times -1 \;\; ; \;\; R_3 \to 3R_3 - R_2 \& R_3 \to {}^{R_3}/_{32} = \begin{bmatrix} 1 & 4 & -1 & |4 \\ 0 & -3 & 10 & |7 \\ 0 & 0 & 1 & |1 \end{bmatrix}$$

$$R_1 \to R_1 + R_3 \& R_2 \to R_2 - 10R_3 = \begin{bmatrix} 1 & 4 & 0 & | \; 5 \\ 0 & -3 & 0 & |-3 \\ 0 & 0 & 1 & | \; 1 \end{bmatrix}$$

$$R_2 \to {}^{R_2}/_{-3} \& R_1 \to R_1 - 4R_2 \begin{bmatrix} 1 & 0 & 0 & |1 \\ 0 & 1 & 0 & |1 \\ 0 & 0 & 1 & |1 \end{bmatrix}$$

Therefore, values of x=1, y=1 and z=1

57. Find the eigen values of $\begin{bmatrix} \mathbf{3} & \mathbf{10} & \mathbf{5} \\ \mathbf{-2} & \mathbf{-3} & \mathbf{-4} \\ \mathbf{3} & \mathbf{5} & \mathbf{7} \end{bmatrix}$

Solution

$$\begin{vmatrix} 3-\lambda & 10 & 5 \\ -2 & -3-\lambda & -4 \\ 3 & 5 & 7-\lambda \end{vmatrix} = 0$$

$$(3-\lambda)[(-3-\lambda)(7-\lambda)+20] - 10[-14+2\lambda+12] + 5(-10+9+3\lambda)$$

$$-63 + 9\lambda - 21\lambda + 3\lambda^2 + 60 + 21\lambda - 3\lambda^2 + 7\lambda^2 - \lambda^3 - 20\lambda + 140$$

$$-20\lambda - 120 - 50 + 45 + 15\lambda = 0$$

$$\lambda^3 - 7\lambda^2 + 16\lambda - 12 = 0\lambda^3 - 7\lambda^2 + 16\lambda - 12 = 0$$

After factorization $\rightarrow (\lambda-2)(\lambda-2)(\lambda-3) = 0$

58. Find rank of the matrix converting into normal form

(a) $\begin{bmatrix} \mathbf{0} & \mathbf{1} & \mathbf{2} & \mathbf{1} \\ \mathbf{1} & \mathbf{2} & \mathbf{3} & \mathbf{2} \\ \mathbf{3} & \mathbf{1} & \mathbf{1} & \mathbf{3} \end{bmatrix}$

Solution

$$R_1 \leftrightarrow R_2 \& R_2 \leftrightarrow R_3 = \begin{bmatrix} 1 & 2 & 3 & 2 \\ 3 & 1 & 1 & 3 \\ 0 & 1 & 2 & 1 \end{bmatrix}$$

$$R_2 \to R_2 - 3R_1 = \begin{bmatrix} 1 & 2 & 3 & 2 \\ 0 & -5 & -8 & -3 \\ 0 & 1 & 2 & 1 \end{bmatrix}$$

$$C_2 \to C_2 - C_4 \& C_3 \to C_3 - 3C_1 = \begin{bmatrix} 1 & 0 & 0 & 2 \\ 0 & -2 & -8 & -3 \\ 0 & 0 & 2 & 1 \end{bmatrix}$$

$$C_4 \to C_4 - 2C_1 \& C_3 \to {}^{C_3}/_2 = \begin{bmatrix} 1 & 0 & 0 & 0 \\ 0 & -2 & -4 & -3 \\ 0 & 0 & 1 & 1 \end{bmatrix}$$

$$R_2 \to R_2 + 3R_3 = \begin{bmatrix} 1 & 0 & 0 & 0 \\ 0 & -2 & -1 & 0 \\ 0 & 0 & 1 & 1 \end{bmatrix}$$

$$R_2 \to R_2 + R_3 = \begin{bmatrix} 1 & 0 & 0 & 0 \\ 0 & -2 & 0 & 1 \\ 0 & 0 & 1 & 1 \end{bmatrix}$$

$$C_4 \to C_4 - C_3 \& C_2 \to C_2 + 2C_4 = \begin{bmatrix} 1 & 0 & 0 & 0 \\ 0 & 0 & 0 & 1 \\ 0 & 0 & 1 & 0 \end{bmatrix}$$

$$C_2 \leftrightarrow C_4 = \begin{bmatrix} 1 & 0 & 0 & 0 \\ 0 & 1 & 0 & 0 \\ 0 & 0 & 1 & 0 \end{bmatrix} = I_3 0$$

Therefore, rank of matrix $\begin{bmatrix} 0 & 1 & 2 & 1 \\ 1 & 2 & 3 & 2 \\ 3 & 1 & 1 & 3 \end{bmatrix}$ is 3

(b) $\begin{bmatrix} \mathbf{1} & \mathbf{1} & \mathbf{1} & \mathbf{1} \\ \mathbf{3} & \mathbf{4} & \mathbf{5} & \mathbf{2} \\ \mathbf{2} & \mathbf{3} & \mathbf{4} & \mathbf{0} \end{bmatrix}$

Solution

$$R_2 \to R_2 - 3R_1 \& R_3 \to R_3 - 2R_1 = \begin{bmatrix} 1 & 1 & 1 & 1 \\ 0 & 1 & 2 & -1 \\ 0 & 1 & 2 & -2 \end{bmatrix}$$

$$R_3 \to R_3 - R_2 = \begin{bmatrix} 1 & 1 & 1 & 1 \\ 0 & 1 & 2 & -1 \\ 0 & 0 & 0 & -1 \end{bmatrix}$$

$$C_3 \to C_3 - C_1 \quad ; \quad C_2 \to C_2 + C_1 \& C_4 \to C_4 - C_1 = \begin{bmatrix} 1 & 0 & 0 & 0 \\ 0 & 1 & 2 & -1 \\ 0 & 0 & 0 & -1 \end{bmatrix}$$

$$C_3 \to C_3 - 2C_2 \& C_4 \to C_4(-1) = \begin{bmatrix} 1 & 0 & 0 & 0 \\ 0 & 1 & 0 & 1 \\ 0 & 0 & 0 & 1 \end{bmatrix}$$

$$C_4 \to C_4 - C_2 = \begin{bmatrix} 1 & 0 & 0 & 0 \\ 0 & 1 & 0 & 0 \\ 0 & 0 & 0 & 1 \end{bmatrix}$$

$$C_3 \leftrightarrow C_4 = \begin{bmatrix} 1 & 0 & 0 & 0 \\ 0 & 1 & 0 & 0 \\ 0 & 0 & 1 & 0 \end{bmatrix}$$

Therefore, rank of matrix $\begin{bmatrix} 1 & 1 & 1 & 1 \\ 3 & 4 & 5 & 2 \\ 2 & 3 & 4 & 0 \end{bmatrix}$ is 3

59. An automobile company uses three types of steel S1, S2 and S3 for producing three types of cars C1, C2 and C3. Steel requirements (in tons) for each type of cars are given below: Using Cramer's rule, find the number of cars of each type which can be produced using 29, 13 and 16 tonnes of steel of three types

	C1	*C2*	*C3*
S1	2	3	4
S2	1	1	2
S3	3	2	1

Solution

Let, Number of cars produced by steel type C1, C2 and C3 be x, y and z respectively

In linear equation system

$2x + 3y + 4z = 29$

$x + y + 2z = 13$

$3x + 2y + z = 16$

Using crammers rule, $\Delta = \begin{bmatrix} 2 & 3 & 4 \\ 1 & 1 & 2 \\ 3 & 2 & 1 \end{bmatrix} = 5$

$$\Delta x = \begin{vmatrix} 29 & 3 & 4 \\ 13 & 1 & 2 \\ 16 & 2 & 1 \end{vmatrix} = 10$$

$$\Delta y = \begin{vmatrix} 2 & 29 & 4 \\ 1 & 13 & 2 \\ 3 & 16 & 1 \end{vmatrix} = 15$$

$$\Delta z = \begin{vmatrix} 2 & 3 & 29 \\ 1 & 1 & 13 \\ 3 & 2 & 16 \end{vmatrix} = 20$$

$$x = \frac{\Delta x}{\Delta} = \frac{10}{5} = 2$$

$$y = \frac{\Delta y}{\Delta} = \frac{15}{5} = 3$$

$$z = \frac{\Delta z}{\Delta} = \frac{20}{5} = 4$$

60. Examine linear dependence or independence of the following set of vectors

(a) $\{[3 \quad 2 \quad 4], [1 \quad 0 \quad 2], [1 \quad -1 \quad -1]\}$

Solution

$$K_1 \begin{bmatrix} 3 \\ 2 \\ 4 \end{bmatrix} + K_2 \begin{bmatrix} 1 \\ 0 \\ 2 \end{bmatrix} + K_3 \begin{bmatrix} 1 \\ -1 \\ -1 \end{bmatrix} = 0$$

$3K_1 + K_2 + K_3 = 0$

$2K_1 + (-K_3) = 0$

$4K_1 + 2K_2 - K_3 = 0$

$$\begin{bmatrix} 3 & 1 & 1 \\ 2 & 0 & -1 \\ 4 & 2 & -1 \end{bmatrix} \begin{bmatrix} K_1 \\ K_2 \\ K_3 \end{bmatrix} = \begin{bmatrix} 0 \\ 0 \\ 0 \end{bmatrix}$$

Augmented matrix, $A : B = \begin{bmatrix} 3 & 1 & 1:0 \\ 2 & 0 & -1:0 \\ 4 & 2 & -1:0 \end{bmatrix}$

$$R_3 \to R_3 - 2R_2 = \begin{bmatrix} 3 & 1 & 1:0 \\ 2 & 0 & -1:0 \\ 0 & 2 & 1:0 \end{bmatrix}$$

$$R_3 \to R_3 + R_2 = \begin{bmatrix} 3 & 1 & 1:0 \\ 2 & 0 & -1:0 \\ 2 & 2 & 0:0 \end{bmatrix}$$

$$R_2 \to R_2 + R_1 \& R_3 \to \tfrac{R_3}{2} = \begin{bmatrix} 3 & 1 & 1:0 \\ 5 & 1 & 0:0 \\ 1 & 1 & 0:0 \end{bmatrix}$$

$$R_3 \to R_3 - R_2 = \begin{bmatrix} 3 & 1 & 1:0 \\ 5 & 1 & 0:0 \\ -4 & 0 & 0:0 \end{bmatrix}$$

$From\ row\ 3 \to\ -4K_1 = 0$

$K_1 = o$

$From\ row\ 2 \to\ 5K_1 + K_2 = 0;\ \ K_2 = 0$

$From\ row\ 1 \to\ 3K_1 + K_2 + K_3 = 0;\ \ K_3 = 0$

Here rank of A is 3 which is equal to number of unknowns and $K_1 = K_2 = K_3 = 0$
Therefore, the vector is linearly independent.

(b) $\{[\mathbf{1}\ \ \mathbf{2}\ \ \mathbf{3}\ \ \mathbf{4}], [\mathbf{0}\ \ \mathbf{1}\ \ \mathbf{-1}\ \ \mathbf{2}], [\mathbf{1}\ \ \mathbf{5}\ \ \mathbf{1}\ \ \mathbf{8}], [\mathbf{3}\ \ \mathbf{7}\ \ \mathbf{8}\ \ \mathbf{14}]\}$

Solution

$$K_1 \begin{bmatrix} 1 \\ 3 \\ 2 \\ 4 \end{bmatrix} + K_2 \begin{bmatrix} 0 \\ 1 \\ -1 \\ 2 \end{bmatrix} + K_3 \begin{bmatrix} 1 \\ 5 \\ 1 \\ 8 \end{bmatrix} + K_4 \begin{bmatrix} 3 \\ 7 \\ 8 \\ 14 \end{bmatrix} = 0$$

$K_1 + 0K_2 + K_3 + 3K_4 = 0$

$2K_1 + K_2 + 5K_3 + 7K_4 = 0$

$3K_1 - K_2 + K_3 + 8K_4 = 0$

$4K_1 + 2K_2 + 8K_3 + 14K_4 = 0$

$$\begin{bmatrix} 1 & 0 & 1 & 3 \\ 2 & 1 & 5 & 7 \\ 3 & -1 & 1 & 8 \\ 4 & 2 & 8 & 14 \end{bmatrix} \begin{bmatrix} K_1 \\ K_2 \\ K_3 \\ K_4 \end{bmatrix} = 0$$

Augmented matrix, $A: B = \begin{bmatrix} 1 & 0 & 1 & 3:0 \\ 2 & 1 & 5 & 7:0 \\ 3 & -1 & 1 & 8:0 \\ 4 & 2 & 8 & 14:0 \end{bmatrix}$

$$R_4 \to R_4 - 4R_1;\ R_3 \to R_3 - 3R_1;\ R_2 \to R_2 - 2R_1 = \begin{bmatrix} 1 & 0 & 1 & 3:0 \\ 0 & 1 & 3 & 1:0 \\ 0 & -1 & -2 & -1:0 \\ 0 & 2 & 4 & 2:0 \end{bmatrix}$$

$$R_4 \to R_4 + 2R_3 = \begin{bmatrix} 1 & 0 & 1 & 3:0 \\ 0 & 1 & 3 & 1:0 \\ 0 & -1 & -2 & -1:0 \\ 0 & 0 & 0 & 0:0 \end{bmatrix}$$

$$R_3 \to R_3 - R_2 = \begin{bmatrix} 1 & 0 & 1 & 3:0 \\ 0 & 1 & 3 & 1:0 \\ 0 & 0 & 1 & 0:0 \\ 0 & 0 & 0 & 0:0 \end{bmatrix}, Rank\ is\ 3$$

Here rank of A is 3 which is less than number of unknowns

(c) $\{[2 \quad -1 \quad 4], [0 \quad 1 \quad 2], [6 \quad -1 \quad 14], [4 \quad 0 \quad 12]\}$

Solution

$$K_1 \begin{bmatrix} 2 \\ -1 \\ 4 \end{bmatrix} + K_2 \begin{bmatrix} 1 \\ 1 \\ 2 \end{bmatrix} + K_3 \begin{bmatrix} 6 \\ -1 \\ 14 \end{bmatrix} + K_4 \begin{bmatrix} 4 \\ 0 \\ 12 \end{bmatrix}$$

$2K_1 + 0K_2 + 6K_3 + 4K_4 = 0$

$-K_1 + K_2 - K_3 + 0K_4 + 0$

$$4K_1 + 2K_2 + 14K_3 + 12K_4 = 0$$

$$\begin{bmatrix} 2 & 0 & 6 & 4 \\ -1 & 1 & -1 & 0 \\ 4 & 2 & 14 & 12 \end{bmatrix} \begin{bmatrix} K_1 \\ K_2 \\ K_3 \\ K_4 \end{bmatrix} = 0$$

$$\text{Augmented matrix, } A:B = \begin{bmatrix} 2 & 0 & 6 & 4:0 \\ -1 & 1 & -1 & 0:0 \\ 4 & 2 & 14 & 12:0 \end{bmatrix}$$

$$R_1 \leftrightarrow R_2 = \begin{bmatrix} -1 & 1 & -1 & 0:0 \\ 2 & 0 & 6 & 4:0 \\ 4 & 2 & 14 & 12:0 \end{bmatrix}$$

$$R_2 \to R_2 + 2R_1 \& R_3 \to R_3 + 4R_1 = \begin{bmatrix} -1 & 1 & -1 & 0:0 \\ 0 & 2 & 4 & 4:0 \\ 0 & 6 & 10 & 12:0 \end{bmatrix}$$

$$R_3 \to R_3 - 3R_2 = \begin{bmatrix} -1 & 1 & -1 & 0:0 \\ 0 & 2 & 4 & 4:0 \\ 0 & 0 & -2 & 0:0 \end{bmatrix}$$

$$-2K_3 = 0; \;\; K_3 = 0$$

$$-K_1 + K_1 - K_3 = 0; \;\; K_1 = K_2$$

$$2K_2 + 4K_3 + 4K_4 = 0; \;\; K_2 = -2K_4$$

Therefore, the vector is linearly independent.

(d) $\{[2 \;\; 3 \;\; -1 \;\; -1], [1 \;\; -1 \;\; -2 \;\; -4], [3 \;\; 1 \;\; 3 \;\; -2], [6 \;\; 3 \;\; 0 \;\; -7]\}$

Solution

$$2K_1 + K_2 + 3K_3 + 6K_4 = 0$$

$$3K_1 - K_2 + K_3 + 3K_4 = 0$$

$$-K_1 - 2K_2 + 3K_3 + 0K_4 = 0$$

$$-K_1 - 4K_2 - 2K_3 - 7K_4 = 0$$

$$\text{Augmented matrix } A:B = \begin{bmatrix} 2 & 1 & 3 & 6:0 \\ 3 & -1 & 1 & 3:0 \\ -1 & -2 & 3 & 0:0 \\ -1 & -4 & -2 & -7:0 \end{bmatrix}$$

$$R_1 \leftrightarrow R_3 = \begin{bmatrix} -1 & -2 & 3 & 0:0 \\ 3 & -1 & 1 & 3:0 \\ 2 & 1 & 3 & 6:0 \\ -1 & -4 & -2 & -7:0 \end{bmatrix}$$

$$R_4 \to R_4 - R_1;\ \ R_3 \to R_3 - 2R_1 \& R_2 \to R_2 + 3R_1 = \begin{bmatrix} -1 & -2 & 3 & 0:0 \\ 0 & -7 & 10 & 3:0 \\ 0 & -3 & 9 & 6:0 \\ 0 & -2 & -5 & -7:0 \end{bmatrix}$$

$$R_3 \to {}^{R_3}/_3 = \begin{bmatrix} -1 & -2 & 3 & 0:0 \\ 0 & -7 & 10 & 3:0 \\ 0 & -1 & 3 & 2:0 \\ 0 & -2 & -5 & -7:0 \end{bmatrix}$$

$$R_4 \to R_4 - 2R_3 \& R_3 \to 7R_3 - R_2 = \begin{bmatrix} -1 & -2 & 3 & 0:0 \\ 0 & -7 & 10 & 3:0 \\ 0 & 0 & 11 & 11:0 \\ 0 & 0 & -11 & -11:0 \end{bmatrix}$$

$$R_4 \to R_4 + R_3 = \begin{bmatrix} -1 & -2 & 3 & 0:0 \\ 0 & -7 & 10 & 3:0 \\ 0 & 0 & 11 & 11:0 \\ 0 & 0 & 0 & 0:0 \end{bmatrix}$$

$K_1 = K_2 = K_3 = K_4 = 0$

Therefore, the vector is linearly dependent.

Chapter 2

Limit and Continuity

1 $lim_{x \to -2} \frac{(x^3+8)}{(x^2-4)}$

$= \frac{(-2^3+8)}{(-2^2-4)}$

$= \frac{-8+8}{4-4}$

$= \frac{0}{0}$ (This is in indeterminate form)

Common indeterminate form $\left(\frac{0}{0}, \frac{\infty}{\infty}\right)$

Applying L-Hospitals rule, we differentiate numerator and denominator separately,

$\frac{\frac{d}{dx}x^3+8}{\frac{d}{dx}x^2-4} = \frac{3x^2}{2x}$

$= \frac{3*4}{-4} = -3$

$$lim_{x\to-2}\frac{x^3+8}{x^2-4}=-3$$

2. $lim_{x\to 0}\frac{1-cosx}{x^2}$

$=\frac{1-cos0}{0}$

$=\frac{1-1}{0}$

=0 (This is in indeterminate form)

Applying L-Hospitals rule, we differentiate numerator and denominator separately,

$\frac{\frac{d}{dx}(1-cosx)}{\frac{d}{dx}(x^2)}=\frac{sinx}{2x}$

$=\lim\limits_{x\to 0}\frac{sinx}{2x}$

Move the term $\frac{1}{2}$ outside of the limit because it is constant with respect to x,

$=\frac{1}{2}\lim\limits_{x\to 0}(\frac{sinx}{x})$

$\frac{d}{dx}=\frac{cosx}{1}$

$=\frac{1}{2}\,lim_{x\to 0}\frac{cos0}{1}$

$=\frac{1}{2}\times 1$

$lim_{x\to 0}\frac{1-cosx}{x^2}=\frac{1}{2}$

3. $lim_{x\to 0}\frac{cos4x-cos6x}{x^2}$

(This is in indeterminate form)

$cos(C)\text{-}cos(D)=2sin\,\frac{C+D}{2}.sin\,\frac{D-C}{2}$

$=2sin\,\frac{4x+6x}{2}.sin\,\frac{6x-4x}{2}$

$=2sin\,\frac{10x}{2}.sin\,\frac{2x}{2}$

$lim_{x\to 0}\frac{(2sin5x.sinx)}{(x^2)}=\frac{2sin5x}{x}.\frac{sinx}{x}$

$lim_{x\to 0}\,2\times 5\,\frac{sinx}{x}.\frac{sinx}{x}=10\frac{sin0}{0}.\frac{sin0}{0}$

$\lim_{x\to 0}\frac{cos4x-cos6x}{x^2}=\mathbf{10}$

4. $lim_{x\to 3}\frac{|x-3|}{x-3}$

LHS: $lim_{x\to 3-}\frac{|x-3|}{x-3}$

As x<3; |x-3|= -(x-3)

$=lim_{x\to 3}\frac{-(x-3)}{x-3}$

$=\frac{-(-6)}{-6}$

= -1

RHS: $lim_{x\to 3+}\frac{|x-3|}{x-3}$

As x>3; |x-3|= x-3

$=lim_{x\to 3+}\frac{x-3}{x-3}$

=1

LHS≠RHS

$lim_{x\to 3}\frac{x-3}{x-3}$ Doesn't exist

5. $lim_{x\to 0}\frac{(1+x)^6-1}{(1+x)^2-1}$

Solution

(This is in indeterminate form)

$lim_{x\to 0}\frac{d}{dx}=lim_{x\to 0}\frac{6(1+x)^5-0}{(1+x)^2-0}$ $\qquad (\frac{(1+x)^5}{(1+x)^1})=(1+x)^{5-1}=(1+x)^4$

$=\lim\limits_{x\to 0} 3(1+x)^4$

$=3(1+0)^4$

$=3$

$$\lim_{x\to 0}\frac{(1+x)^6-1}{(1+x)^2-1}=3$$

6. $lim_{x\to 0}\frac{sin7x}{tan8x}$

Solution

$lim_{x\to 0}\frac{\frac{sin7x}{7x}\times 7x}{\frac{tan8x}{8x}\times 8x}$

$=lim_{x\to 0}\frac{\frac{sin7x}{7x}\times 7}{\frac{tan8x}{8x}\times 8}$

$=\frac{(1)7}{(1)8}=\frac{7}{8}$

$$\lim_{x\to 0}\frac{sin7x}{tan8x}=\frac{7}{8}$$

7. $\lim_{x\to\infty} \frac{x^2+x+1}{x^3+2x^2-3x}$

Solution

(This is in indeterminate form) dividing num/den by highest power of x i.e.,) x^3

$$=\lim_{x\to\infty} \frac{\left(\frac{x^2+x+1}{x^3}\right)}{\left(\frac{x^3+2x^2-3x}{x^3}\right)}$$

$$=\lim_{x\to\infty} \frac{\left(\frac{1}{x}+\frac{1}{x^2}+\frac{1}{x^3}\right)}{\left(1+\frac{2}{x}-\frac{3}{x^2}\right)} = \frac{0}{1}$$

$$\lim_{x\to\infty} \frac{x^2+x+1}{x^3+2x^2-3x} = 0$$

8. If f(x) =$\frac{x}{|x|}$ where x≠0

f(x) =1 where x=0

Check the continuity of f(x) at x=0

Solution

LHL =$\lim_{x\to 0-} \frac{x}{|x|} = \frac{x}{-x}$= -1

RHL =$\lim_{x\to 0+} \frac{x}{|x|} = \frac{x}{x}$=1

f(x)=1 at x=0

here, LHL≠RHL

Then f(x) =$\frac{x}{|x|}$is not continuous at x=0

9. Find the value of k for which $\mathbf{f(x)=\frac{sin5x}{3x}}$ **if x≠0**

=k if x=0

is continuous at x=0

Solution

LHL:

$$lim_{x\to 0-}\frac{sin5x}{3x}=\frac{1}{3}.\frac{sin5x}{x}$$ Using L-hospital Rule

$$lim_{x\to 0-}\frac{5}{3}.\frac{sinx}{x}=\frac{5}{3}lim_{x\to 0}\frac{sinx}{x}$$

$$\frac{5}{3}\times 1=\frac{5}{3}$$

RHL

$$lim_{x\to 0+}\frac{sin5x}{3x}.=\frac{5}{3}$$

As f(x)=k if x=0

Value of K= $\frac{5}{3}$

10. Let $\boldsymbol{f(x)=3x-4}$ **when 0≤x≤2**

$\boldsymbol{=2x+\beta}$ **when 2<x≤3**

If f(x) is continuous at x=2, then β=?

Solution

LHL= $lim_{x\to 2-}3\text{x}-4$

=3(2)-4

=6-4

=2

RHL= $lim_{x\to 2+}2\text{x}+\beta$

=4+β

f(x) is continuous at x=2

so, LHS = RHS

4+β=2

Value of β=-2

11. Let $f(x)=kx^2$, if x≤2

=3 if kx^2x>2

If f(x) is continuous at x=2 then value of k

Solution

f(x) is continuous at 2

kx^2=3 at x=2 $\Rightarrow lim_{x\to 2^-} kx^2 = lim_{x\to 2^+} 3$

$k(2)^2=3$

k(4)=3

value of k=$\frac{3}{4}$

12. $f(x) = 1 + x \text{ when } x \leq 2$
$= 5 - x \text{ when } x > 2$

Check differentiability at $x = 2$

Solution

LHD =>x<2

$f'(2) = lim_{x\to 2^-} \frac{f(x)-f(2)}{x-2}$ (here; x=2-h)

$= lim_{h\to 0} \frac{f(2-h)-f(2)}{2-h-2}$

$= lim_{h\to 0} \frac{f(2-h)-f(2)}{-h}$

$= lim_{h\to 0} \frac{1+2-h-(1+2)}{-h}$

$= lim_{h\to 0} \frac{3-h-3}{-h}$

$= \frac{-h}{-h}$

=1

In case of RHD =x>2

f|(2) = $lim_{x\to 2+}\frac{f(x)-f(2)}{x-2}$

= $lim_{h\to 0}\frac{f(2+h)-f(2)}{2+h-2}$

= $lim_{h\to 0}\frac{5-(2+h)-(5-2)}{h}$

= $lim_{h\to 0}\frac{3-h-3}{h}$

= $\frac{-h}{h}$

=-1

LHD≠RHD

Therefore, the function is not differentiable at x=2

13. f(x)=$\frac{x-1}{2x^2-7x+5}$, x $\neq$ 1 x is not equal to 1

=$\frac{-1}{3}$ when x=1

Check differentiation at x=1

Solution

f(x)= $\frac{x-1}{2x^2-7x+5}$

$\frac{x-1}{2x^2-2x+5x+5}$

$\frac{x-1}{2x(x-1)-5(x-1)}$

$\frac{x-1}{(2x-5)(x-1)}$

f(x)=$\frac{1}{2x-5}$

f(1)=$\frac{1}{2-5}$

f(1)=$\frac{-1}{3}$

Differentiability:

$f|(x) = \frac{f(x+h)-f(x)}{h}$

$f|(1) = \frac{f(1+h)-f(1)}{h}$

LHD

$f(x-0) = lim_{h\to 0-} \frac{\left(\frac{1}{2(1+h-5)}\right)+\frac{1}{3}f(1)}{h}$

$lim_{h\to 0-} \frac{\frac{1}{2+2h-5}+\frac{1}{3}}{h}$

$lim_{h\to 0} \frac{\frac{1}{2h-3}+\frac{1}{3}}{h}$

$= \lim_{h\to 0} \frac{3+2h-3}{(2h-3)3*h}$

$= \frac{2h}{3(2h-3)h}$

$= \frac{-2}{9}$

RHD

$f'(1+0) = lim_{x\to 0} \frac{(1+h)-f(1)}{h}$

$\frac{lim_{h\to 0}\frac{1}{2(1+h)-5}+\frac{1}{3}}{h}$

$\frac{-2}{9}$

LHD = RHD

Therefore, the function is differentiable at x=1.

14. $\boldsymbol{If\ y = tan(x+y), find\ \frac{dy}{dx}}$.

Solution

Differentiation with respect to 'x' on both sides

$\frac{d(y)}{dx} = \frac{d(tan\,(x+y))}{dx}$

$dy/dx = sec^2(x+y) \ * \ (1 + dy/dx)$

$dy/dx = sec^2(x+y) \ + \ sec^2(x+y)dy/dx$

$dy/dx - sec^2\,(x+y)\,dy/dx \ = \ sec^2\,(x+y)$

$\left(\frac{dy}{dx}\right) sec^2\,(x+y) \ = \ sec^2(x+y)$

$dy/dx = sec^2(x+y)/((1 - sec^2(x+y))$

$sec^2\,(x+y)/(1 - [1 + tan^2\,(x+y)]\,)$

$(1 + tan^2(x+y))/(1 - 1 - tan^2\,(x+y)\,)$

$(1 + tan^2(x+y))/(-tan^2(x+y)\,)$

$(-1)/tan^2(x+y)$ - $tan^2\,(x+y)/tan^2\,(x+y)$

$(-1)/tan^2(x+y)$ - 1

$-cot^2(x+y) - 1$

$-[cot^2(x+y) + 1]$

$dy/dx = \boldsymbol{-cosec^2\,(x+y)}$

15. $\boldsymbol{y = \sqrt{sinx} \ + \ \sqrt{sinx} \ + \sqrt{sinx} \ + \sqrt{sinx} \ + \cdots .. \infty \ find \ \frac{dy}{dx}}$

Solution

$y = \sqrt{sinx} + y$

$y^2 = sin\ x + y$

$2y.\frac{dy}{dx} \ = \ cos\ x + \frac{dy}{dx}$

$2y.\frac{dy}{dx} - \frac{dy}{dx} \ = \ cosx$

$\frac{dy}{dx}(2y - 1) = cosx$

$$\frac{dy}{dx} = \frac{cosx}{2y - 1}$$

16. $\boldsymbol{x = 2t/(1 + t^2)}$

$\boldsymbol{y = (1 - t^2)/(1 + t^2) find\ dy/dx}$

Solution

Let, $t = tan\theta$

$\theta = tan^{-1}t$

$$x = \frac{2tan\theta}{(1 + tan^2\theta)} = sin2\theta$$

$y = (1 - tan^2\theta)/(1 + tan^2\theta) = cos2\theta$

$y = 1 - tan^2\theta/1 + tan^2\theta = cos2\theta$

$dx/d\theta = cos2\theta \times 2$ $dy/d\theta = -sin2\theta \times 2$

$dy/dx = -2\ sin2\theta/2\ cos2\theta$

$-tan2\theta$

$$= \frac{-2\tan\theta}{1 - \tan^2\theta}$$

$$= -\left(\frac{2t}{1 - t^2}\right)$$

Chapter 3

Differentiation

1. $f(x) = 1 + x$ **when x≤2**

   $= 5 - x$ **when x>2**

   Check differentiability at x=2

Solution

In LHD, x < 2

L[f'(2^-)] = $lim_{x\to 2} \frac{f(x)-f(2)}{x-2}$ (here; x=2-h)

$= lim_{h\to 0} \frac{f(2-h)-f(2)}{2-h-2}$

$= lim_{h\to 0} \frac{f(2-h)-f(2)}{-h}$

$= lim_{h\to 0} \frac{1+2-h-(1+2)}{-h}$

$= lim_{h\to 0} \frac{3-h-3}{-h}$

$= \frac{-h}{-h} = 1$

R[f'(2)]=$lim_{x\to 2+}\frac{f(x)-f(2)}{x-2}$

$=lim_{h\to 0}\frac{f(2+h)-f(2)}{2+h-2}$

$=lim_{h\to 0}\frac{5-(2+h)-(5-2)}{h}$

$=lim_{h\to 0}\frac{3-h-3}{h}$

$=\frac{-h}{h}$

=-1

LHD≠RHD

Therefore, the function is not differentiable at x=2

2. $f(x) = 2 - x$ **when x≤2**

$= x - 2$ **when x>2**

Check differentiability at x=2

Solution

LHD = x<2

L[f'(2)] =$lim_{x\to 2}\frac{f(x)-f(2)}{x-2}$

$=lim_{h\to 0}\frac{f(2-h)-f(2)}{2-h-2}$

$=lim_{h\to 0}\frac{2-(2-h)-(2-2)}{-h}$

$=\frac{2-2+h}{-h}$

= -1

RHD: x>2

R[f'(2)]=$lim_{x\to 2+}\frac{f(x)-f(2)}{x-2}$

$$=lim_{h\to 0}\frac{f(2+h)-f(2)}{2+h-2}$$

$$=lim_{h\to 0}\frac{(2+h-2)-(2-2)}{h}$$

$$=\frac{h}{h}$$
$$=1$$

LHD≠RHD

Therefore, the function is not differentiable @x=2

3. **Check differentiability of $f(x) = |x - 1|$ at x=1**

Solution

f(x)=|x-1|= {x-1 if x>1

{-(x-1) if x<1

LHD: x<1

$$L[f^{\prime}(1)] = lim_{x\to 1-}\frac{f(x)-f(1)}{x-1}$$

$$=lim_{h\to 0}\frac{f(1-h)-f(1)}{1-h-1}$$

$$=lim_{h\to 0}\frac{-(1-h-1)-[-(1-1)]}{-h}$$

$$=\frac{-1+h+1}{-h}$$

=-1

RHD: x>1

$$R[f^{\prime}(1)]=lim_{x\to 1+}\frac{f(x)-f(1)}{x-1}$$

$$=lim_{h\to 0}\frac{f(1+h)-f(1)}{1+h-1}$$

$$=lim_{h\to 0}\frac{1+h-1-(1-1)}{h}$$

$=\frac{h}{h}$

$=1$

LHD$\neq$RHD

Therefore, the function is not differentiable @x=1

4. $f(x) = x^2$ **if x$\geq$ 0**

$= x$ **if x<0**

Check differentiability at x=0

Solution

LHD =x<0

$Lf^{l}(0) = lim_{x\to 0-} \frac{f(x)-f(0)}{x-0}$

$= lim_{h\to 0} \frac{f(0-h)-f(0)}{0-h-0}$

$= lim_{h\to 0} \frac{0-h-0}{-h}$

$=1$

RHD: x>0

$R[f^{l}(2)] = lim_{x\to 0+} \frac{f(x)-f(0)}{x-0}$

$= lim_{h\to 0} \frac{f(0+h)-f(0)}{0+h-2}$

$= \frac{lim_{h\to 0}(0+h)^2-0}{h}$

$=0$

LHD$\neq$RHD

Therefore, the function is not differentiable @x=0

5. $f(x) = \frac{x-1}{2x^2-7x+5}, x \neq 1$

$= \frac{-1}{3}$ **when x=1**

Check differentiation at x=1

Solution

$$f(x) = \frac{x-1}{2x^2-7x+5}$$

$= \frac{x-1}{2x^2-2x+5x+5}$

$= \frac{x-1}{2x(x-1)-5(x-1)}$

$= \frac{x-1}{(2x-5)(x-1)}$

$f(x) = \frac{1}{2x-5}$

$f(1) = \frac{1}{2-5}$

$f(1) = \frac{-1}{3}$

Differentiability:

$f^{l}(x) = \frac{f(x+h)-f(x)}{h}$

$f^{l}(1) = \frac{f(1+h)-f(1)}{h}$

LHD:

$$f(x\text{-}0) = lim_{x\to 0-} \frac{f\left(\frac{1}{2(1+h-5)}\right)+\frac{1}{3}f(1)}{h}$$

$$= lim_{x\to 0-} \frac{f\left(\frac{1}{2+2h-5}\right)+\frac{1}{3}}{h}$$

$$= lim_{x\to 0} \frac{\frac{1}{2h-3}+\frac{1}{3}}{h}$$

$$= \frac{3+2h-3}{(2h-3)3\times h}$$

$$= \frac{2h}{3(2h-3)h}$$

$$= \frac{-2}{9}$$

RHD:

$$f^{I}(1+0) = lim_{x\to 0} \frac{(1+h)-f(1)}{h}$$

$$= \frac{lim_{x\to 0}\frac{1}{2(1+h)-5}+\frac{1}{3}}{h}$$

$$= \frac{-2}{9}$$

LHD = RHD

DIFFERENTIATION OF IMPLICIT FUNCTIONS

Therefore, the function is differentiable at x=1

1. If $xy = x^3+y^3$, find $\frac{dy}{dx}$

Solution

Differentiation with respect to 'x' on both sides

$$\frac{d(xy)}{dx} = \frac{d}{dx}(x^3+y^3)$$

$$y\frac{d(x)}{dx}+x\frac{d(y)}{dx}=\frac{d}{dx}(x^3)+\frac{d}{dx}y^3$$

$$y+x\frac{dy}{dx}=3x^2+3y^2\frac{dy}{dx}$$

$$x\frac{dy}{dx}-3y^2\frac{dy}{dx}=3x^2-y$$

$$\frac{dy}{dx}=\frac{3x^2-y}{x-3y^2}$$

2. **If $y = tan(x + y)$, find $\frac{dy}{dx}$.**

Solution

Differentiation with respect to 'x' on both sides

$$\frac{d(y)}{dx}=\frac{d}{dx}(\tan(x+y))$$

$$\frac{dy}{dx}=\sec^2 x+y\,(1+\frac{dy}{dx})$$

$$\frac{dy}{dx}=\sec^2 x+y+\sec^2(x+y)\frac{dy}{dx}$$

$$\frac{dy}{dx}-\sec^2(x+y)\frac{dy}{dx}=\sec^2(x+y)$$

$$\frac{dy}{dx}\sec^2(x+y))=\sec^2(x+y)$$

$$\frac{dy}{dx}=\frac{sec^2(x+y)}{(1-sec^2(x+y))}$$

$$=\frac{sec^2(x+y)}{1-[1+tan^2(x+y)]}$$

$$=\frac{1+tan^2(x+y)}{1-1-tan^2(x+y)}$$

$$=\frac{1+tan^2(x+y)}{-tan^2(x+y)}$$

$$=\frac{-1}{tan^2(x+y)}-\frac{tan^2(x+y)}{tan^2(x+y)}$$

$$=\frac{-1}{tan^2(x+y)}-1$$

$=-cot^2(x+y)-1$

$=-[cot^2(x+y)+1]$

$$\frac{dy}{dx}=\mathbf{-cosec^2(x+y)}$$

3. $y=\sqrt{sinx+\sqrt{sinx+\sqrt{sinx+\sqrt{sinx}}}}+.....\infty$

Solution

$y=\sqrt{sinx+y}$

$y^2=\sin x+y$

$$2y.\frac{dy}{dx}=\cos x+\frac{dy}{dx}$$

$$2y.\frac{dy}{dx}-\frac{dy}{dx}=\cos x$$

$$\frac{dy}{dx}(2y-1)=\cos x$$

$$\frac{dy}{dx}=\frac{cosx}{2y-1}$$

4. $x=y+\cfrac{1}{y+\cfrac{1}{y+\cfrac{1}{y+\cfrac{1}{y+...\infty}}}}$ **prove that** $\frac{dy}{dx}=2x^2+y^2-3xy$

Solution

$$\frac{d}{dx}(x)=\frac{dy}{dx}+\frac{d}{dx}(x^{-1})$$

$1=\frac{dy}{dx}+(-1)\,(x^{-1})^{-1}$

$\frac{dy}{dx}=1+x^{-2}$

$=1+1/x^2$

$=1+\left(\frac{1}{x}\right)^2$

$=1+(x-y)^2$ As $x=y+1/x$

$\frac{dy}{dx}=1+x^2+y^2-2xy$ ->1 Equation 1

$X=y+\frac{1}{x}$

$X=\frac{xy+1}{x}$

$X^2=xy+1$

$1=x^2-xy$

From Equation 1

$\frac{dy}{dx}=x^2-xy+x^2+y^2-2xy$

$\frac{dy}{dx}=\mathbf{2x^2+y^2-3xy}$

Hence proved.

5. If sin y = x(sin(x+y)), then prove that $\frac{dy}{dx}=\sin^2\frac{(a+y)}{\sin a}$

Solution

$\sin y = x.\sin(x+y)$

$\cos y\frac{dy}{dx}=\sin(a+y).1+x.\cos(a+y).(0+\frac{dy}{dx})$

$$\cos y\frac{dy}{dx} = \sin(a+y) + x.\cos(a+y).\frac{dy}{dx}$$

$$\cos y\frac{dy}{dx} - x\cos(a+y)\frac{dy}{dx} = \sin(a+y)$$

$$\frac{dy}{dx}[\cos y - x\cos(a+y)] = \sin(a+y)$$

$$\frac{dy}{dx} = \frac{\sin(a+y)}{\cos y - x\cos(a+y)}$$

$$= \frac{\sin(a+y)}{\cos y - \frac{\sin y}{\sin(a+y)}.\cos(a+y)} \text{ [x = siny/sin(a+y)]}$$

$$= \frac{\sin(a+y)}{\frac{\cos y.\sin(a+y) - \sin y.\cos(a+y)}{\sin(a+y)}}$$

$$=\sin^2(a+y)/\cos y.\sin(a+y) - \sin y.\cos(a+y)$$

$\frac{dy}{dx}$**$=\sin^2(a+y)/\sin a$ [cosy.sin(a+y)-siny.cos(a+y)=sin(a+y-y)=sina]**

Hence proved.

PARAMETRIC EQUATION

1. **x=a cosθ**

   y=a sinθ find dy/dx

Solution

x=a cosθ

d.w.r to θ

$$\frac{dx}{d\theta} = -a\sin\theta$$

y=asinθ

$$\frac{dy}{d\theta} = a\cos\theta$$

$$\frac{dy}{dx} = \frac{dy/_{d\theta}}{dx/_{d\theta}} = \frac{acos\theta}{-asin\theta} = -cot\theta$$

2. $\boldsymbol{x = \frac{2t}{1+t^2}}$

$\boldsymbol{y = \frac{1-t^2}{1+t^2}}$ **find** $\boldsymbol{\frac{dy}{dx}}$

Solution

let t=tanθ

θ=tan⁻¹t

$$x = \frac{2\tan\theta}{1+tan^2\theta} = sin2\theta \; y = \frac{1-tan^2\theta}{1+tan^2\theta} = cos2\theta$$

$$\frac{dx}{d\theta} = 2cos2\theta \; \frac{dy}{d\theta} = -2sin2\theta$$

$$\frac{dy}{dx} = \frac{-2sin2\theta}{2cos2\theta}$$

$$= -tan2\theta$$

$$= \frac{-2tan\theta}{1 - tan^2\theta}$$

$$= -\frac{-2t}{1 - t^2}$$

3. **Find dc of tanx w.r.t cotx**

Solution

(d(tan x)/dx)/(d(cot x)/dx)

$$= \frac{sec^2 x}{-cosec^2 x}$$

$$= \frac{{}^{-1}/_{cos^2x}}{{}^{1}/_{sin^2x}}$$

$$= \frac{-sin^2x}{cos^2x}$$

$$= -tan^2x$$

4. Find d.c of cosx w.r.t x³

Solution

$$\frac{dcosx}{dx^3}$$

$$= \frac{-sinx}{3x^2}$$

7. Find d.c of $tan^{-1}\left(\frac{2x}{1-x^2}\right)$ w.r.t to $\cos^{-1}\left(\frac{1-x^2}{1+x^2}\right)$

Solution

let x=tanθ

θ=tan⁻¹x

$$= \frac{\tan^{-1}(2x)}{1-x^2}$$

$$= \frac{\tan^{-1}(2tan\theta)}{(1-tan^2\theta)}$$

$$=\tan^{-1} tan2\theta$$

=2θ

Also, $\cos^{-1}\left(\frac{1-x^2}{1+x^2}\right)$

$$= \frac{\cos^{-1}(1-tan^2\theta)}{1+tan^2\theta}$$

$=\cos^{-1}\cos 2\theta$

$=2\theta$, required differentiation

$$= \frac{d2\theta}{d2\theta} = 1$$

6. **If** $y = e^{x+e^{x+e^{x+\cdots\infty}}}$ **prove that** $\frac{dy}{dx} = \frac{y}{1-y}$

Solution

$$y = e^{x+y}$$

$$\frac{dy}{dx} = \frac{d(e^{x+y})}{dx}$$

$$e^{x+y}\left(1+\frac{dy}{dx}\right)$$

$$\frac{dy}{dx} = e^{x+y} + e^{x+y}\frac{dy}{dx}$$

$$\frac{dy}{dx} - e^{x+y}\frac{dy}{dx} = e^{x+y}$$

$$\frac{dy}{dx}(1-e^{x+y}) = e^{x+y}$$

$$\frac{dy}{dx} = \frac{e^{x+y}}{1-e^{x+y}}$$

$$\frac{dy}{dx} = \frac{y}{1-y}$$

Hence proved.

7. If x=logt + sint

y=e^t + logt, $\frac{dy}{dx} = ?$

Solution

x=logt+sint

$$\frac{dx}{dt} = \frac{1}{t} + cost$$

$$y = e^t + cost$$

$$\frac{dy}{dt} = e^t - sint$$

$$\frac{dy}{dx} = \frac{dy/_{dt}}{dx/_{dt}} = \frac{e^t - sint}{1/_t + cost}$$

LOGARITHM DIFFERENTIATION

1. If x^y=y^{x}finddy/dx

Solution

log on both sides

$$=logx^y = logy^x$$

$$=ylogx = xlogy$$

d.w.t x (uv format)

$$=logx\frac{dy}{dx} + \frac{y}{x} = logy + \frac{x}{y}\frac{dy}{dx}$$

$$=logx\frac{dy}{dx} - \frac{x}{y}\frac{dy}{dx} = logy - \frac{x}{y}$$

$$=\frac{dy}{dx} = \frac{logy - \frac{y}{x}}{logx - \frac{x}{y}}$$

2. **if $x^y = e^{x-y}$ find $\frac{dy}{dx}$**

Solution

$$ylogx = (x - y)loge$$

$$ylogx = x - y$$

$$\frac{y}{x} = \frac{1}{logx + 1}, y\ logx + y = x$$

Differentiation wrt x on both sides

$$\frac{dy}{dx}(logx + 1) + \frac{y}{x} = 1$$

$$\frac{dy}{dx} = \frac{1-\frac{y}{x}}{logx+1}$$

$$\frac{dy}{dx} = \frac{1-\frac{1}{logx+1}}{logx+1} = \frac{logx+1-1}{logx+1} \times \frac{1}{logx+1}$$

$$\frac{logx}{(logx+1)^2}$$

3. **If $y = x^{\frac{1}{x}}$ find $\frac{dy}{dx}$=?**

Solution

log on both sides

$$logy = \frac{1}{x}logx$$

d.w.r to x

$$\frac{d(logy)}{dx} = \frac{d(x^{-1}logx)}{dx}$$

$$\frac{1}{y}\frac{dy}{dx} = logx \times (-1)x^{-2} + x^{-1} \times \frac{1}{x}$$

$$\frac{dy}{dx} = y\left(\frac{1}{x^2} - \frac{logx}{x^2}\right)$$

$$\frac{dy}{dx} = \frac{y}{x^2}(1 - logx)$$

APPROXIMATION VALUE

1. $(1.999)^6$

Formula; $f'(a) = [f(a+h) - f(a)]/h$

$$f(a+h) = hf'(a) + f(a)$$

Solution

1.999 = 2+ (-0.001); a=2, h=-0.001

$f(x) = x^6(a+h) = 1.999$

$f'(x) = 6x^5 f(x) = 2^6$

$f'(a) = 6a^5$ $\qquad f(2) = 64$

$f^1(2) = 6\times2^5$

$= 6\times32$

$f^1(2) = 192$

$f(a+h) = hf'(a) + f(a)$

$f(1.999) = (-0.001)(192) + 64$

=-0.192+64

$\mathbf{f(1.999) = 63.808}$

2. $(33)^{1/5}$

Solution

33 = 32+1; a=32, h=1

$f(x) = x^{1/5}$ $\quad$ a + h = 33

$f'^{(x)} = \frac{1}{5}x^{4/5}$ $\quad f(32) = 32^{1/5}$

$f'(a) = \frac{1}{5} a^{4/5} \qquad f(32) = 2^{5\times\frac{1}{5}}$

$f'(32) = \frac{1}{5} \times 32^{4/5} \qquad f(32) = 2$

$=1/5\times2^{5\times4/5}$

$=1/5\times16$

$=1/80$

$f'(32) = 1/80$

$f(a+h) = hf'(a) + f(a)$

$f(33) = \frac{1}{80} + 2$

$\frac{161}{80} = 2.0125$

$\mathbf{f(33) = 2.0125}$

3. $(63)^{1/3}$

Solution

63 = 64+(-1); a=64, h=-1

$f(x)=x^{1/3}$ a+h=63

$f'(x)=1/3x^{2/3}$ $f(64)=64^{1/3}$

$f'(a)=1/3\times a^{2/3}$ $f(64)=4^{3*1/3}$

$f'(64)= 1/3\times64^{2/3}$ f(64)=4

$=1/3\times4^{3\times2/3}$

$=1/3\times4^2$

$=1/3\times16$

$f'(32)=1/48$

$f(a+h) = hf'(a) + f(a)$

$f(63) = (-1)\frac{1}{48} + 4$

$\frac{191}{48} = 3.97$ f (63) =3.97

4. $(0.999)^5$

Solution

0.999 = 1+(-0.001); a=1, h=-0.001

$F(x) = x^5 \quad a+h=0.999$

$f^1(x) = 5x^4 \qquad f(a)=a^5$

$f^1(a) = 5\times a^4 \qquad f(1) = 1^5$

$f^1(1) = 5\times 1^4 \qquad f(1) = 1$

=5

$f^1(1)=5$

$$f(a+h) = hf'(a) + f(a)$$

$$f(0.999) = (-0.001)\times 5 + 1$$

=1-0.005

f (0.999) =0.995

5. $(124)^{1/3}$

Solution

124 = 125+(-1); a=125, h=-1

$f(x) = x^{1/3} \quad a+h=124$

$f^1(x)=1/3x^{2/3} \qquad f(125) = 125^{1/3}$

$f^1(a)=1/3\times a^{2/3} \qquad f(125) = 5^{3\times 1/3}$

$f^1(125) = 1/3\times 125^{2/3} \qquad f(125) = 5$

$=1/3\times 5^{3\times 2/3}$

$=1/3\times 5^2$

$=1/3\times 25$

$f^1(125)=1/75$

$$f(a+h) = hf'(a) + f(a)$$

$$f(124) = 5 + (-1)\frac{1}{75}$$

=374/75

f(124)=4.98

Chapter 4

Maxima and Minima Values

- ☆ Differentiate the given function.
- ☆ Let, $f'(x) = 0$ and find critical numbers
- ☆ Then find the second derivative $f''(x)$.
- ☆ Apply those critical numbers in the second derivative.
- ☆ The function $f(x)$ is maxima when $f''(x) < 0$
- ☆ The function $f(x)$ is minima when $f''(x) > 0$
- ☆ To find the maximum and minimum value we need to apply those x values in the given function.

1. If $y = f(x) = 2x^3 - 15x^2 + 36x + 1$

Solution

$$y = 2x^3 - 15x^2 + 36x + 1$$

$$\frac{dy}{dx} = 6x^2 - 30x + 36$$

for maxima and minima

$$\frac{dy}{dx} = 0$$

$$6x^2 - 30x + 36 = 0$$

$$x^2 - 5x + 6 = 0$$

$$x^2 - 3x - 2x + 6 = 0$$

$$(x - 3)(x - 2) = 0$$

x=3,2

$$\frac{d\frac{dy}{dx}}{dx} = \frac{d^2y}{dx^2} = 12x - 30$$

when x=3

$$\frac{d^2y}{dx^2} = 12 \times 3 - 30 = 6$$

when x=2

$$\frac{d^2y}{dx^2} = 12 \times 2 - 30 = -6$$

when $\frac{d^2y}{dx^2}$ at x=2 is less than 0 the function

$\boldsymbol{y = f(x) = 2x^3 - 15x^2 + 36x + 1}$ has maxima value at x=2

max value=$2\times2^3-15(2)^2+36\times2+1$

= 29

$\frac{d^2y}{dx^2}$ at x=3>0 the function y=fx has minimum value at x=3

Min value=28

2. Find the maxima & minima value of $y = f(x) = x^{\frac{1}{x}}$

Solution

$$y = x^{\frac{1}{x}}$$

$$logy = \frac{logx}{x}$$

$$\frac{1}{y}\frac{dy}{dx} = \frac{1}{x^2} + logx(-1)(x)^{-2}$$

$$\frac{1}{y}\frac{dy}{dx} = \frac{1}{x^2} - \frac{logx}{x^2}$$

$$\frac{dy}{dx} = y\left(\frac{1}{x^2} - \frac{logx}{x^2}\right)$$

for maxima and minima $\frac{dy}{dx}$=0

$$y\left(\frac{1}{x^2} - \frac{logx}{x^2}\right) = 0$$

$$\frac{1 - logx}{x^2} = 0$$

$$1 - logx = 0$$

$$logx = 1$$

x=e

x=2.71

as x>2 it has minima value

value=$x^{\frac{1}{x}}$

=$2.71^{1/2.71}$

=1.45

3. **Find the maxima and minima of $y = x^3 - 2x^2 + x + 6$**

Solution

$$y = x^3 - 2x^2 + x + 6$$

$$\frac{dy}{dx} = 3x^2 - 4x + 1$$

for maxima and minima

$$\frac{dy}{dx} = 0$$

$$3x^2 - 4x + 1 = 0$$

$$3x^2 - x - 3x + 1 = 0$$

$$x(3x - 1) - 1(3x - 1) = 0$$

$$(x - 1)(3x - 1) = 0$$

x=1 or x=1/3

$$\frac{d^2y}{dx^2} = 6x - 4$$

when x=1

$$\frac{d^2y}{dx^2}=2$$

when x=1/3

$$\frac{d^2y}{dx^2}=-2$$

As $\frac{d^2y}{dx^2}$ at x=1/3 is less than zero the function has maxima value at x=1/3

max value is $y = \frac{1}{3^3} - \frac{2}{3^2} + \frac{1}{3} + 6$

=6.148

min value = $1^3 - 2^2 + 1 +$

=6

4. Maxima and minima of $y = \frac{logx}{x}$

Solution

$$\frac{dy}{dx} = \frac{x\frac{d(logx)}{dx} - logx\frac{dx}{dx}}{x^2}$$

$$\frac{x \times \frac{1}{x} - logx}{x^2}$$

$$\frac{1 - logx}{x^2}$$

$$\frac{dy}{dx} = 0$$

$$\frac{1 - logx}{x^2} = 0$$

$$1 - logx = 0$$

$$logx = 1$$

x=10

$$\frac{d^2y}{dx^2} = \frac{d\,\frac{1-logx}{x^2}}{dx}$$

$$\frac{x^2\left(1 - \frac{1}{x}\right) - (1 - logx)2x}{x^4}$$

$$\frac{-x - 2x + 2xlogx}{x^4}$$

$$\frac{x(2logx) - 3}{x^4}$$

$$\frac{2logx - 3}{x^3}$$

for x=10

$$\frac{d^2y}{dx^2} = \frac{2log10 - 3}{10^3} = \frac{2 - 3}{1000} = \frac{-1}{1000}$$

As, $\frac{d^2y}{dx^2} < 0$ at x=10 so, we have maximum value at x=10

Maximum value is

$$y = \frac{logx}{x} = \frac{log10}{10} = \frac{1}{10} = 0.1$$

5. What is the value of the function (x – 1) (x – 2)² at its maxima?

Solution

Given,

$$f(x) = (x - 1)(x - 2)^2$$

$$f(x) = (x - 1)(x^2 + 4 - 4x);$$

$$f(x) = (x^3 - 5x^2 + 8x - 4)$$

Now, $f'(x) = 3x^2 - 10x + 8,\ f'(x) = 0$

$$3x^2 - 10x + 8 = 0$$

$$(3x - 4)(x - 2) = 0$$

$$x = \frac{4}{3}, 2$$

Now, $f''(x) = 6x - 10$

$$f''\left(\frac{4}{3}\right) = 6 \times [4/3] - 10 < 0$$

$$f''(2) = 12 - 10 > 0$$

Hence, at x = 4 / 3 the function will occupy maximum value.

∴ Maximum value = $f\left(\frac{4}{3}\right) = \frac{4}{27}$

6. Find the maximum and minimum value of the function
$\boldsymbol{x^3 - 3x^2 - 9x + 12}$

Solution

Let $y = f(x) = x^3 - 3x^2 - 9x + 12$

$f'(x) = 3x^2 - 3(2x) - 9(1) + 0$

$f'(x) = 3x^2 - 6x - 9$

$f'(x) = 0$

$3x^2 - 6x - 9 = 0$

÷ by 3 => $x^2 - 2x - 3 = 0$

x + 1= 0

x = -1,

x - 3 = 0

x = 3.

$f'(x) = 3x^2 - 6x - 9$

$f''(x) = 3(2x) - 6(1) - 0$

$f''(x) = 6x - 6$

Put x = -1

$f''(-1) = 6(-1) - 6$

$-6 - 6$

f''(-1) = -12 < 0, Maximum

To find the maximum value let us apply x = -1 in the given function

$f(x) = x^3 - 3x^2 - 9x + 12$

Solution

$f(-1) = (-1)^3 - 3(-1)^2 - 9(-1) + 12$

=-1 - 3(1) + 9 + 12

-1 - 3 + 9 + 12

= -4 + 21

= 17

Put x = 3

$f''(3) = 6(3) - 6$

18 - 6

f''(3) = 12 < 0, Minimum

To find the minimum value let us apply x = 3 in the given function

$f(x) = x^3 - 3x^2 - 9x + 12$

Solution

$f(3) = 3^3 - 3(3)^2 - 9(3) + 12$

= 27 - 3(9) - 27 + 12

= 27 - 27 - 27 + 12

= -27 + 12

= -15

Therefore, the maximum value = 17 and the minimum value = -15.

7. **Find the maximum and minimum value of the function**
   $\mathbf{4x^3 - 18x^2 + 24x - 7}$

Solution

Let, $y = f(x) = 4x^3 - 18x^2 + 24x - 7$

$f'(x) = 4(3x^2) - 18(2x) + 24(1) - 0$

$f'(x) = 12x^2 - 36x + 24$

$f'(x) = 0$

$12x^2 - 36x + 24 = 0$

÷ by 12 => $x^2 - 3x + 2 = 0$

x - 1 = 0　　　　x - 2 = 0

x = 1　　　　x = 2

$f'(x) = 12x^2 - 36x + 24$

$f''(x) = 12(2x) - 36(1) + 0$

$f''(x) = 24x - 36$

Put x = 1

$$f''(1) = 24(1) - 36$$

$$24 - 36$$

$$f''(1) = -12 < 0 \; Maximum$$

8. To find the maximum value let us apply x = 1 in the given function.

$$f(x) = 4x^3 - 18x^2 + 24x - 7$$

$$f(1) = 4(1)^3 - 18(1)^2 + 24(1) - 7$$

$$= 4(1) - 18(1) + 24 - 7$$

$$= 4 - 18 + 24 - 7$$

$$= 28 - 25$$

$$= 3$$

Put x = 2

$$f''(2) = 24(2) - 36$$

$$= 48 - 36$$

$$f''(2) = 12 > 0 \; Minimum$$

9. To find the minimum value let us apply x = 2 in the given function

$$f(x) = 4x^3 - 18x^2 + 24x - 7$$

$$f(2) = 4(2)^3 - 18(2)^2 + 24(2) - 7$$

= 4(8) - 18(4) + 48 - 7

= 32 - 72 + 48 - 7

= 80 - 79

= 1, Therefore the maximum value is 3 and the minimum value is 1.

10. The sum of two positive numbers is equal to 'a' and if the sum of their cubes is least, the numbers are?

Solution

Let one of the number be 'x'. Then another number will be (a−x). Hence,

=> $f(x) = x^3 + (a - x)^3$

$Differentiating\ with\ respect\ to\ x, gives\ us$

$=> f'(x) = 3x^2 - 3(a-x)^2$

$=> f'(x) = 0 \ implies$

$=> x^2 - (a-x)^2 = 0$

$=> (2x-a)(a) = 0$

$Since \ a = 0$

$Hence,$

$x = 2a.$

$Therefore \ both \ the \ numbers \ are$

$2a, 2a.$

11. Find two positive numbers x and y such that $x + y = 60 \ and \ xy^3$ is maximum.

Solution

Let, $P = xy^3$

It is given that x+y=60

$\Rightarrow x = 60 - y$

$P = (60-y)\ y^3$ $[Putting \ value \ of \ x]$

$= 60y^3 - y^4$

$\Rightarrow \frac{dy}{dP} = 180y^2 - 4y^3$

$dy^2/d^2P = 360y - 12y^2$

For maximum or minimum values of y, P we have

$dy/dP = 0$

$\Rightarrow 180y^2 - 4y^3 = 0$

$\Rightarrow 4y^2(45-y) = 0$

$\Rightarrow$ y=0 45−y=0 **y=45**

Now, $(dy^2/d^2P)\ y = 45 = 360 \times 45 - 12(45)^2$

=12×45−(30−45)

$=-8100<0$

P is maximum when y=45

When, y=45, $x+y=60 \Rightarrow x=60-45$

x=15

Numbers are 15 and 45.

12. Find local maxima and local minima for the function f(x) =x³−3x.

Solution

$f(x) = x^3 - 3x.$

$f'(x) = 3x^2 - 3 = 0$

$\rightarrow x^2 - 1 = 0$

$\rightarrow x = \pm 1.$

$f''(x) = 6x$

$Hence\ f''(-1) < 0\ maxima.$

$f''(1) > 0\ is\ minima.$

Hence f(x) attains the maximum value at x=−1 and a minimum value at x=1 f(−1)=3−1=2. The maximum value is 2.

13. Show that $f(x) = x^x$ is minimum when $= e^{-1}$.

Solution

Step 1: Apply Natural Log on both sides

Given,

$f(x) = (x^x)$

Applying natural log on both sides we get

$ln\ f(x) = ln(x^x)$

$ln\ f(x) = x\ ln\ x$

Step 2: Differentiate the equation with respect to x,

Differentiating, we get

$$f'(x)\frac{1}{f\,(x)} = ln\ x + \frac{1}{x}$$

$$f'(x) = f(x)(lnx + 1)$$

$$But,\ f(x) = x^x$$

$$\therefore f'(x) = x^x(lnx + 1)$$

Step 3: Find value of x where f(x) is maximum

Equate f'(x) to 0

$$f'(x) = 0$$

$$x^x(lnx + 1)\ = 0$$

Here, x^x cannot be zero, hence lnx+1=0

$$ln\ x = -1\ \Rightarrow\ x = \frac{1}{e}$$

Hence, it has been proven that, the given function f(x) is minimum, when x = $\frac{1}{e}$

GREATEST AND LEAST VALUES

Working Rule

1. In order to find the greatest or least value of f(x) in in interval, write down the sign scheme of f'(x) in that interval. For this equate f'(x) =0 and from this find the values of x. These values of x will divide the given interval in various sub interval.
2. f(x) will be an increasing or decreasing function in an interval according as f'(x) >0 or <0 in that interval
3. f(x) will have local minima and local maxima in interval [a, b] at x=a and x=b respectively if it is an increasing function in [a,b]
4. f(x) will have local maxima and local minima in interval [a, b] at x=a and x=b respectively if it is an decreasing function in [a,b]
5. If f(x) is continuous in the given interval then greatest values of f(x) will be the greatest of all local maxima and least value of f(x) will be the least of all minima.

1. **Find the Greatest and Least values of the function, $f(x) = x^4 - 2x^2 + 5$ in the interval [-2,2]**

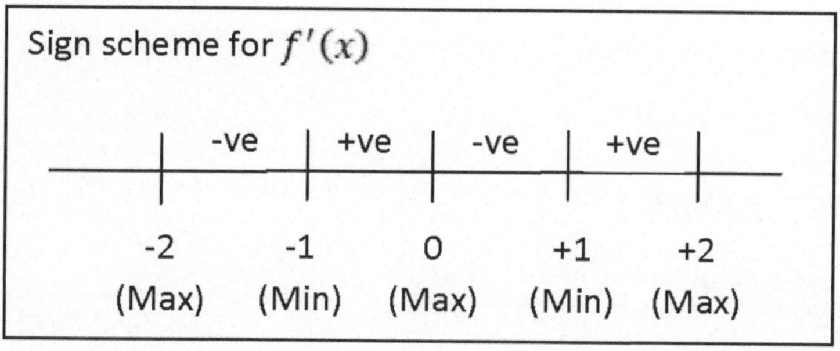

Solution

$$f(x) = x^4 - 2x^2 + 5$$

$$f'(x) = 4x^3 - 4x$$

$f'(x) = 0 = 4x^3$-4x=0

$4x(x^2 - 1)$=0

x=0; x=1; x=-1

As f'(x) at x=-2 is equal to -24, Also f'(x) <0 in [-2,-1] so decreasing function in given interval(sign scheme shown above)

Maximum values are found at x=-2,x=0,x=2

Maxima value at x=-2

=>$(-2)^4$-2$(-2)^2$+5

=16-8+5

=13

Maxima value at x=0=>5

Maxima value at x=2=>13

Greatest value is 13

Minimum values are found at x=-1 and x=1

Minima value at x=-1=>1-2+5=4

Minima value at x=1=>1-2+5=4

Least value=4

2. Find the Greatest and Least square of function, f(x)=x^3-3x^2+6x-2 in the interval [-1, 1]

Solution

f(x) =x^3-3x^2+6x-2

=3x^2-3(2X)+6

=3x^2-6x+6

f'(x)=0=>3x^2-6x+6=0

x^2-2x+2=0

$$x = \frac{-b \pm \sqrt{b^2 - 4ac}}{2a}$$

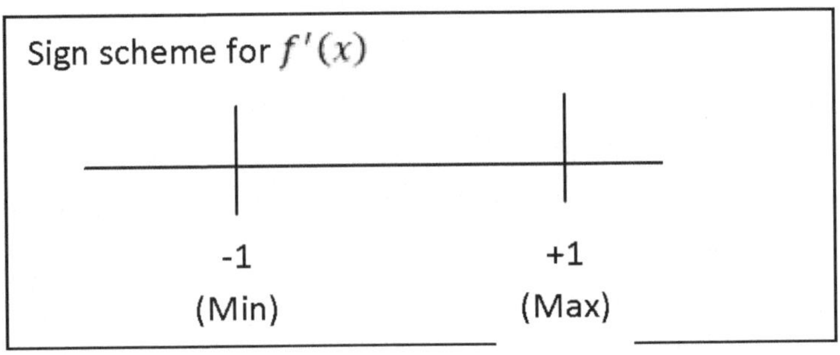

$$= \frac{-(-2) \pm \sqrt{-(2)^2 - 4*1*2}}{2*1}$$

$$= \frac{2 \pm \sqrt{-4}}{2}$$

$$= \frac{2 \pm \sqrt{4i^2}}{2}$$

$$= \frac{2 \pm 2i}{2} = \frac{2(1 \pm i)}{2} = 1 \pm i$$

X= 1+i and 1-i

i.e., roots are imaginary also f'(x) is greater than zero *i.e.,* $3x^2 - 6x + 6 > 0$ for all values of X, thus f'(x) > 0 in [-1,1]

Also, at x=0, fI(x)=0-0+6= +6 (>0)

As f'(x) at x=0 is +ve if it is an increasing function.

Therefore f(x) is an increasing function in [-1,1] (sign scheme for f'(x) is shown above).

Function has maximum value at x=1 and function have minimum value at x= -1

Maximum value is $f(x) = 1^3 - 3 \times (-1)^2 + 6 \times (-1) - 2$

$$= -1 - 3 - 6 - 2 = -12$$

3. Find the Greatest and Least values of the function,f(x)=$\sin 2x - x$, in the interval of $[\frac{-\pi}{2}, \frac{\pi}{2}]$

Solution

f'(x)=$\cos 2x$×2-1

=2$\cos 2x$-1

f'(x)=0 =>2$\cos 2x$-1=0

$2\cos 2x = 1$

$$\cos 2x = \frac{1}{2}$$

$\cos 2x = \cos 60^0$

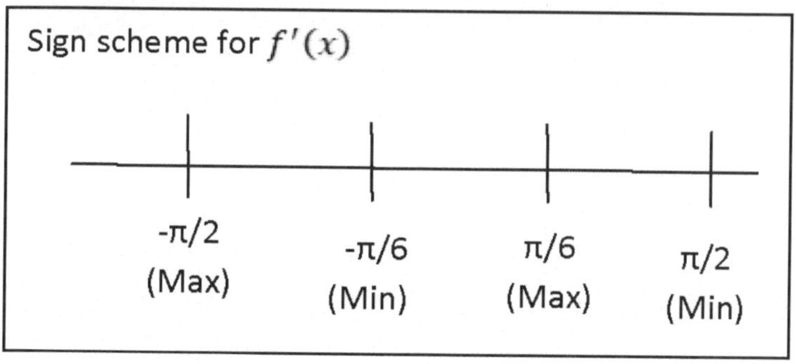

$2x=60^0$

$2x=\frac{\pi}{3}$

$X=\frac{\pi}{6}$

Also, as $\cos 2x = \frac{1}{2} = \cos\frac{\pi}{3} = cos - \left(\frac{\pi}{3}\right)$

Therefore, $2x = -\frac{\pi}{3}; x = -\frac{\pi}{6}$

As interval is defined in $[\frac{-\pi}{2}, \frac{\pi}{2}]$,so sign scheme of f'(x) i.e., $2cos2x - 1$ is given above.

So, f(x) has maximum value at x= -π/2 & π/6

Maxima values at $x=\frac{-\pi}{2}$

Maximum value $=sin2\left(\frac{-\pi}{2}\right) - \left(\frac{-\pi}{2}\right)$

$= -\sin\pi + \left(\frac{\pi}{2}\right) = 0 + \frac{\pi}{2} = \frac{\pi}{2}$

Maxima values at $x=\frac{\pi}{6}$

Maximum value $=sin2\left(\frac{\pi}{6}\right) - \left(\frac{\pi}{6}\right)$

$=sin\,\frac{\pi}{3} - \frac{\pi}{6}$

$=\frac{\sqrt{3}}{2} - \frac{\pi}{6}$

Greatest value of f(x) in $\left[\frac{-\pi}{2}, \frac{\pi}{2}\right] = \frac{\pi}{2}$

Minima value of f(x)at x= $-\frac{\pi}{6}$

Minimum value of f(x) $=sin2\left(-\frac{\pi}{6}\right)-\left(-\frac{\pi}{6}\right)$

$=sin\left(\frac{-\pi}{3}\right)+\frac{\pi}{6}$

$$=-\frac{\sqrt{3}}{2}+\frac{\pi}{6}$$

Minima value of f(x) at $x=\frac{\pi}{2}$

Minimum value $=sin2\left(\frac{\pi}{2}\right)-\frac{\pi}{2}$

$=-\frac{\pi}{2}$

So, least value of f(x) in $\left[\frac{-\pi}{2},\frac{\pi}{2}\right]=-\frac{\pi}{2}$

DIFFERENTIATION – WORD PROBLEMS

1. A spherical balloon is pumped at the rate of 10cu.inches/min. Find the rate of increase of radius when its radius is 15 inches.

Solution

Volume of a sphere, $v=\frac{4}{3}\pi r^3$

Take v as y and r as x

$$y=\frac{4}{3}\pi x^3$$

$\frac{dy}{dt}=\frac{4}{3}\pi\frac{dx3}{dt}$

$=\frac{4}{3}\pi\times 3x^2\frac{dx}{dt}$

$\frac{dy}{dt}=4\pi x^2\frac{dx}{dt}$

Given, $\frac{dy}{dt}$ =10 inches at r=15(x=15)

$$10=4\times\pi\times(15\times15)\frac{dx}{dt}$$

$$\frac{dx}{dt}=\frac{70}{4\times22\times15\times15}=\frac{70}{19800}$$

$$=0.0035\text{inches/min}$$

Rate of increase of radius = 0.0035inches/min

2. **If the radius of circle increases at a uniform rate of 6cm/sec. Find the rate of increase of area when radius is 50cm.**

Solution

$$\frac{dr}{dt}=6\text{cm/sec}$$

Area of circle, $A = \pi r^2$

Diff. with respect to t,

$$\frac{dA}{dt}=\pi\frac{dr2}{dt}\times\frac{dr}{dt}$$

$$=\pi\times2r.\frac{dr}{dt}$$

$$= \pi\times2r\times6$$

$$= \pi\times2(50)\times6$$

$$=1886.71\text{cm}^3/\text{sec}$$

Rate of increase of area=1886.71cm³/sec

3. **If the side of an equilateral triangle increases uniformly at the rate of 3ft/sec, at what rate is the area increasing when side is 10ft.**

Solution

Area of equilateral triangle, $A = \frac{\sqrt{3}}{4}a^2$

$\frac{dA}{dt}=\frac{\sqrt{3}}{4}\frac{da2}{dt}.\frac{da}{dt}$

$=\frac{\sqrt{3}}{4}.2a.\frac{da}{dt}$

$=\frac{\sqrt{3}}{4}.2(10).3$

$=25.98$

Rate of increase of area = 25. 98sq.ft/sec

4. A metal cube of side 3cm is heated. Find the approximate in its volume, when its side becomes 3.0001cm.

Solution

Volume of cube,V = a^3

Change in side of cube, $\frac{da}{dt}$ = 3.0001-3; a=3.0001

=0.0001cm

$\frac{dv}{dt}=3a^2\frac{da}{dt}$ $=3\times(3.0001)^2.0.0001$

$=0.0027$

Approximate value of volume = 0.0027cu.cm

5. Find the approximation error in x^3 corresponding to small error of r% in value of x.

Solution

Let $y=x^3$

$\frac{dy}{dx}=3x^2\frac{dx}{dx}$

$=3x^2$

$\frac{dy}{dx}=\frac{\lambda y}{\lambda x}$

$\lambda y=\frac{dy}{dx}\times\lambda x=3x^2\times\frac{r}{100}\times x=3x^3\times\frac{r}{100}$

Change in y = λy

For a change in y, $\frac{\lambda y}{y}$

For 100, $\frac{\lambda y}{y}\times 100$

$$=\frac{3x\frac{3\times r}{100}*100}{y}$$

$$=\frac{3.x3\times r}{x3} \qquad =3r$$

Approximation error in x^3 = 3r

Chapter 5

Taylor's and Maclauri Theorem

1. Expand sinx as an infinite series

Solution

Let $f(x) = \sin x$

$$f(x) = f(0) + xf'(0) + \frac{x^2}{2!}f''(0) + \frac{x^3}{3!}f'''(0) + \cdots$$

Here in x tend to be zero

$f(x) = \sin x \; f(0) = \sin 0 = 0$

$f'(x) = \cos x \; f'(0) = \cos 0 = 1$

$f''(x) = -\sin x \; f''(0) = -\sin 0 = 0$

$f'''(x) = -\cos x \; f'''(0) = -\cos 0 = -1$ and so on

from the formula:

$$sin\ x = \frac{x}{1!} - \frac{x^3}{3!} + \frac{x^5}{5!} + \frac{x^7}{7!} + \frac{x^9}{9!} + \cdots$$

$$cos\ x = 1 - \frac{x^2}{2!} + \frac{x^4}{4!} + \frac{x^6}{6!} + \cdots$$

2. **Determine the Taylor series at X=0 for f(x) = e^x**

Solution

Given f(x)= e^x

Differentiate the given equation,

$f'(x) = e^x$

$f''(x) = e^x$

$f'''(x) = e^x$

at x= 0, we get

$f'(0) = e^0 = 1$

$f''(0) = e^0 = 1$

$f'''(0) = e^0 = 1$

when Taylor series at x= 0, then Maclaurin series is

$$f(x)= f(0) + f'(0)x + \frac{f''(0)}{2!}x^2 + \frac{f'''(0)}{3!}x^2 + \ldots\ldots\ldots\ldots\ldots$$

$$e^x = 1 + x(1) + (\frac{x^2}{2!})(1) + (\frac{x^3}{3!})(1) + \ldots\ldots\ldots\ldots$$

$$\text{Therefore, } e^x = 1 + x + (\frac{x^2}{2!}) + (\frac{x^3}{3!}) + \ldots\ldots\ldots\ldots$$

3. **Evaluate the Taylor series for f(x) = cos (x) for x=0**

Solution

We need to take the derivatives of the cos x and evaluate them at x=0

$f(x) = \cos x \Rightarrow f(0) = 1$

$f'(x) = -\sin x \Rightarrow f'(0) = 0$

$f''(x) = -\cos x \Rightarrow f''(0) = -1$

$f'''(x) = \sin x \Rightarrow f'''(0) = 0$

$f(5)(x) = -\sin x \Rightarrow f(5)(0) = 0$

$f(6)(x) = -\cos x \Rightarrow f(6)(0) = -1$

Therefore, according to the Taylor series expansion;

$$\text{Cos } x = \Sigma \frac{f^n(0)}{n!} x^n$$

$$= f(0) + f'(0)x + \frac{f''(0)}{2!}x^2 + \frac{f'''(0)}{31}x^3 + \frac{f''''(0)}{4!}x^4 + \frac{f'''''(0)}{5!}x^5$$

$$= 1 + 0 - \frac{1}{2!}x^2 + 0 + \frac{1}{4!}x^4 + \frac{1}{6!}x^6 + \ldots\ldots$$

$$\text{Cos x} = 1 - \frac{1}{2!}x^2 + \frac{1}{4!}x^4 - \frac{1}{6!}x^6 + \ldots\ldots$$

$$\text{Cos x} = \sum_{n=0}^{\infty} \frac{(-1)^n}{(2n)!} x^{2n}$$

4. Expand e^{-x}

Solution

$f(x) = e^{-x}$ $\qquad f(0) = e^0 = 1$

$f'(x) = -e^{-x}$ $\qquad f'(0) = -1$

$f''(x) = e^{-x}$ $\qquad f''(0) = 1$

$f'''(x) = -e^{-x}$ $\qquad f'''(0) = -1$

$$f(x) = f(0) + \frac{x}{1!}f'(0) + \frac{x^2}{2!}f''(0) + \frac{x^3}{3!}f'''(0) + \cdots$$

$$= 1 + x(-1) + \frac{x^2}{2!}(1) + \frac{x^3}{3!}(-1) + \cdots$$

$$e^{-x} = 1 - x + \frac{x^2}{2!} - \frac{x^3}{3!}$$

5. Expand sinx in the power of $(x - \frac{\pi}{2})$

Solution

$$\sin\left(\frac{\pi}{2} + x - \frac{\pi}{2}\right)$$

Assume $a = \frac{\pi}{2}$ and $h = x - \frac{\pi}{2}$

$$f(a+h) = f(a) + fhf'^{(a)} + \frac{h}{2!}f''(a) + \frac{h}{3!}f'''(a) + \cdots$$

$$f(x) = \sin x$$

$$f(a) = \sin a$$

$$f\left(\frac{\pi}{2}\right) = \sin\frac{\pi}{2} = 1$$

$$f'\left(\frac{\pi}{2}\right) = \cos\frac{\pi}{2} = 0$$

$$f''\left(\frac{\pi}{2}\right) = -\sin\frac{\pi}{2}$$

$$f\left(\frac{\pi}{2} + x - \frac{\pi}{2}\right) = 1 + \left(x - \frac{\pi}{2}\right)\cos\frac{\pi}{2} + \frac{\left(x - \frac{\pi}{2}\right)^2}{2}\left(-\sin\frac{\pi}{2}\right) + \cdots$$

$$= 1 + \left(x - \frac{\pi}{2}\right) \times 0 - \frac{1}{2}\left(x - \frac{\pi}{2}\right)2 + \cdots$$

$$= 1 - \frac{1}{2}\left(x - \frac{\pi}{2}\right)2 + \cdots$$

6. Expand $\log \sin (x + h)$ by taylors theorem

Solution

$$f(x+h) = \log\sin(x+h)$$

Let, $f(x) = \log\sin x$

$$f'(x) = \frac{\cos x}{\sin x} = \cot x$$

$$f''(x) = -cosec^2 x$$

$$f'''(x) = -2\ cosec\ x \times -cosec\ x \cot c$$

$$= 2\ cosed^2 x \cot x$$

$$f(x+h) = \log\sin x + h\cot x + h^2 \frac{(-cosec^2 x)}{2}$$

$$+ h^3 \frac{3\ cosec\ 2x \cot x}{6}$$

$$= \log\sin x + h\cot x - h^2 \frac{cosec^2 x}{2} + h^3 \frac{cosec^2 x \cot x}{3} + \cdots$$

7. **Evaluate the Taylor series for f (x) = x³ -10x² +6 at x=3**

Solution: first, we will find the derivatives of the given function

$f(x) = x^3 - 10x^2 + 6 \Rightarrow f(3) = -57$

$f'(x) = 3x^2 - 20x \Rightarrow f'(3) = 3$

$f''(x) = 6x - 20 \Rightarrow f''(3) = 6$

$f'''(x) = 6 \Rightarrow f'''(3) = -2$

$f''''(x) = 0$

Therefore, the required series is:

$$X^3 - 10x^2 + 6 = \sum_{n=0}^{\infty} \frac{f^n(3)}{n!}(x-3)^n$$

$$= f(3) + f'(3)(x-3) + \frac{f''(3)}{2!}(x-3)2 + \frac{f'''(3)}{3!}(x-3)^3 + 0$$

$$= -57 - 33(x-3) - (x-3)^2 + (x-3)^3.$$

7. **Expand 2x³ +7x ² + x –6**

Solution

Let $f(x) = 2x^3 + 7x^2 + x - 6$

Let $f(x) = f[2+(x-2)]$

By Taylor theorem:

$f(x) = f[2+(x-2)] = f(2) + (x-2)f'(2)(x-2)/2!\ f''(2) + (x-2)/3!\ f'''(2) + \ldots\ldots\ldots\ldots$

$f(x) = 2x^3 + 7x^2 + x - 6 \Rightarrow f(2) = 40$

$f'(x) = 6x^2 + 14x + 1 \Rightarrow f'(2) = 53$

$f''(x) = 12x + 14 \Rightarrow f''(2) = 38$

$f'''(x) = 12 \Rightarrow f'''(2) = 12$

$f''''(x) = 0 \Rightarrow f''''(2) = 0$

$$\therefore f(x) = 40 + (x-2)(53) + \frac{(x-2)}{21}(38) + \frac{(x-2)^2}{3!} + \frac{(x-3)^3}{3!}(12) + \ldots\ldots\ldots$$

$f(x) = 40 + 53(x-2) + 19(x-2)^2 + 2(x-2)^3$

8. Use Taylor theorem to find $\sqrt{25.15}$

Solution

let $f(a+h) = \sqrt{25.15} = \sqrt{25+0.15}$

Here a =25 h = 0.15

∴ By Taylor theorem

$$f(a+h) = f(a) + h f'(a) + \frac{h^2}{2!} f''(a) + \frac{h^3}{3!} f'''(a) \ldots\ldots.$$

$\therefore f(a) = \sqrt{25} = \sqrt{a} = 5$

$f'(x) = 1/2\sqrt{a}$

$$\therefore f(a+h) = \sqrt{25.15} = \sqrt{25 + 0.15} = 5 + \frac{0.15}{2}(5) - \frac{(0.15)^2}{2(25)^{\frac{3}{2}}} + \frac{3(0.15)^3}{48(25)^{\frac{5}{8}}}$$

= 5 + 0.015 – 0.000225 + 0.000000675

= 5.01478.

9. Find the first 4 terms of the Taylor series for the following functions:

(a) ln x centered at a=1, (b) $\frac{1}{x}$ centered at a=1, (c) sin x centered at a = $\frac{\pi}{4}$.

Solution

(a) f (x) = ln x. So, $f'(x) = \frac{1}{x} f^{(2)}(x) = \frac{-1}{x^2}$, $f^{(3)}(x) = \frac{2}{x^3}$, $f^{(4)}(x) = -\frac{6}{x^4}$ and so

$$\ln x = \ln 1 + (x-1)\times 1 + \frac{(x-1)^2}{2!}\times(-1) + \frac{(x-1)^3}{3!}\times(2) + \frac{(x-1)^4}{4!}\times(-6) + \ldots$$

$$= (x-1) - \frac{(x-1)^2}{2!} + \frac{(x-1)^3}{3!} - \frac{(x-1)^4}{4!} + \ldots$$

(b) $f(x) = \frac{1}{x}$. So, $f^{(1)}(x) = \frac{1}{x^2}$, $f^{(2)}(x) = \frac{1}{x^3}$, $f^{(3)}(x) = -\frac{6}{x^4}$ and so

$$\frac{1}{x} = 1 + (x-1)\times(-1) + \frac{(x-1)^2}{2!}\times(2) + \frac{(x-1)^3}{3!}\times(-6) + \cdots$$

$$= 1 - (x-1) + (x-1)^2 - (x-1)^3 - \cdots$$

(c) $f(x) = \sin x$. So, $f^{(1)}(x) = \cos x$, $f^{(2)}(x) = -\sin x$, $f^{(3)}(x) = -\cos x$ and so

$$\sin x = \frac{\sqrt{2}}{2} + (x-\frac{\pi}{4})\times\frac{\sqrt{2}}{2} + (x-\frac{\frac{\pi}{4}}{2!})^2 - \frac{\sqrt{2}}{2} + \frac{(x-\pi/4)^3}{3!}\times(-\frac{\sqrt{2}}{2}) + \ldots\ldots$$

$$= \frac{\sqrt{2}}{2}\left[1 + (x-\frac{\pi}{4}) - \frac{(x-\frac{\pi}{4})^2}{2} - \frac{(x-\frac{\pi}{4})^3}{6} + \ldots\right].$$

10. Find the Taylor series for the function $x^4 + x - 2$ centered at a=1.

Solution

$f(x) = x^4 + x - 2$, $f'(x) = 4x^3 + 1$, $f''(x) = 12x^2$, $f'''(x) = 24x$, $f''''(x) = 24$ and all other derivatives are zero.

Thus

$$x^4 + x - 2 = 0 + (x-1)\times 5 + \frac{(x-1)^2}{2!}\times 12 + \frac{(x-1)^3}{3!}\times 24 + \frac{(x-1)^4}{4!}\times 24$$

$$= 5(x-1) + 6(x-1)^2 + 4(x-1)^3 + (x-1)^4$$

LAGRANGE'S MEAN VALUE THEOREM

If f(x) is continuous in [a,b] and differentiable in]a,b[, then there exist a point 'c' in between [a,b], such that,

$$f'(c) = \frac{f(b) - f(a)}{b - a}$$

11. Let $f(x) = 3x^2 + 5x + 7$ be a function defined on [1,3] then the point c €]1,3[such that $f^1(c)=0$

Solution

$$f(1) = 3 + 5 + 7 = 15$$

$$f(3) = 49$$

$$f'(x) = 6x + 5$$

$$f'(c) = \frac{f(b) - f(a)}{b - a} = \frac{49 - 15}{3 - 1} = 1$$

$$6x + 5 = 17$$

$$6c = 17 - 5$$

$$c = 2$$

2. Verify rolle's theorem for $f(x) = 2x^3 + x^2 - 4x - 2$ where $-1/2 \le x \le \sqrt{2}$

Solution

Clearly f(x) is finite for all x and hence f(x) is differentiable at all x, therefore f(x) is continuous at all x and hence continuous in [-1/2, $\sqrt{2}$].

$$f\left(-{}^{1}/_{2}\right) = 2\left(-\tfrac{1}{8}\right) + \tfrac{1}{4} - 4\left(-\tfrac{1}{2}\right) - 2 = 0$$

$$f(\sqrt{2}) = 2(\sqrt{2})^3 + (\sqrt{2})^2 - 4(\sqrt{2}) - 2 = 0$$

$$f\left(-\tfrac{1}{2}\right) = f(\sqrt{2}) = 0$$

Hence verified

3. Let $f(x) = x^3 - 6x^2 + 11x - 6$ be a function defined on [1,3] then the point c€]1,3[such that f¹(c)=0 is given by

Solution

$$f(x) = x^3 - 6x^2 + 11x - 6$$

$$f(1) = 1 - 6 + 11 - 6 = 12$$

$$f(3) = 27 - 54 + 33 - 6 = 0$$

$$f'(x) = 3x^2 - 12x + 11$$

$$f'(c) = 3c^2 - 12c + 11$$

$$0 = 3c^2 - 12c + 11$$

$$c = \frac{-b \pm \sqrt{(b^2 - 4ac)}}{2a}$$

$$= \frac{-(-12) \pm \sqrt{((-12)^2 - 4 \times 3 \times 11)}}{2 \times 3}$$

$$= \frac{12 \pm \sqrt{(144 - 132)}}{6}$$

$$= \frac{12 \pm \sqrt{12}}{6}$$

$$= 2 \pm \frac{\sqrt{12}}{6}$$

$$= 2 \pm \frac{1}{\sqrt{3}}$$

4. **If Rolle's theorem holds for $f(x) = x^3 - 6x^2 + kx + 5$ on [1,3] with $c = 2 + \frac{1}{\sqrt{3}}$, find the value of k.**

Solution

$$f(x) = x^3 - 6x^2 + kx + 5$$

$$f'(x) = 3x^2 - 12x + k$$

$$f'(c) = 3c^2 - 12c + k$$

$f^1(c)=0$(Rolle's theorem)

$$3c^2 - 12c + k = 0$$

Again, $c = 2 + \frac{1}{\sqrt{3}}$ (given)

$$3(2 + \frac{1}{\sqrt{3}})^2 - 12(2 + \frac{1}{\sqrt{3}}) + k = 0$$

$$3(4 + \frac{4}{\sqrt{3}} + {}^1/_3) \quad - 12(2 + \frac{1}{\sqrt{3}}) + k = 0$$

$$13 - 24 + k = 0$$

$$k = 11$$

5. **If $f(x) = x(x-1)(x-2)$ be defined in [0,1/2] then find the value of c of mean value theorem.**

Solution

$$f(x) = x(x-1)(x-2)$$

$$f(x) = x(x^2 - 2x - 2x + 2)$$

$$f(x) = x^3 - 3x^2 + 2x$$

$$f'(x) = 3x^2 - 6x + 2$$

$$f'(c) = 3c^2 - 6c + 2$$

$$f'(c) = \frac{f(b) - f(a)}{b - a}$$

$$f'(c) = \frac{f\left(\frac{1}{2}\right) - f(0)}{\frac{1}{2} - 0}$$

$$3c^2 - 6c + 2 = \frac{f\left(\frac{1}{2}\right) - f(0)}{\frac{1}{2}}$$

$$f\left(\frac{1}{2}\right) = (\frac{1}{2})^3 - 3\left(\frac{1}{2}\right)^2 + 2\left(\frac{1}{2}\right) = \frac{3}{8}$$

$$f(0) = 0$$

$$3c^2 - 6c + 2 = \frac{\frac{3}{8} - 0}{\frac{1}{2}}$$

$$3c^2 - 6c + 2 = \frac{3}{4}$$

$$12c^2 - 24c + 8 = 3$$

$$12c^2 - 24c + 5 = 0$$

$$c = \frac{-b \pm \sqrt{(b^2 - 4ac)}}{2a}$$

$$= \frac{-(-24) \pm \sqrt{(-24^2 - 4 \times 12 \times 5)}}{2 \times 12}$$

$$= \frac{24 \pm \sqrt{(336)}}{24} = 1 \pm \frac{\sqrt{21}}{6}$$

6. **Find the Maclaurin series for ln (1+ x) and hence that for $\ln(\frac{1+x}{1-x})\frac{1+x}{1-x})$**

Solution

The Maclaurin series for ln(1+ x) is standard:

$\ln(1+x) = x-\frac{x^2}{2}+\frac{x^3}{3}-\ldots\ldots$ and so, $\ln(1-x) = -x-\frac{x^2}{2}-\frac{x^3}{3}\ldots\ldots$

Hence $\ln\left(\frac{1+x}{1-x}\right) = \ln(1+x) - \ln(1-x) = 2\left(x+\frac{x^3}{3}+\frac{x^5}{5}+\ldots\right)$

$$2\sum_{n=1}^{\infty}\frac{x^{2n-1}}{2n-1}.$$

7. **(a) Evaluate the integral $\int_0^{\frac{\pi}{6}}\sin^2 x \, dx$ by first finding the Maclaurin approximation to the integrand with 3 terms.**

   (b) Evaluate the integral exactly and compare.

Solution

(a) The Maclaurin series for $\sin^2 x = x^2 - \frac{x^4}{3} + \frac{2x^6}{45}$*.Hence*

$$\int_0^{\frac{\pi}{6}} \sin^2 x \, dx = \int_0^{\frac{\pi}{6}}\left(x^2 - \frac{x^4}{3} + \frac{2x^6}{45}\right)dx = \left(\frac{x^3}{3}-\frac{x^5}{15}+\frac{2x^7}{315}\right) = 0.0452941.$$

b) The exact integral is

$$\int_0^{\frac{\pi}{6}} \sin^2 x \, dx = \int_0^{\frac{\pi}{6}}\frac{1-\cos 2x}{2}dx = \left(\frac{x}{2}-\frac{\sin 2x}{4}\right)\left[\frac{x}{2}-\frac{\sin 2x}{4}\right]_0^{\frac{\pi}{6}}$$

$$= \frac{\pi}{12} - \frac{\sin\left(\frac{\pi}{3}\right)}{4}.$$

$$= 2\pi - \frac{3\sqrt{3}}{24} = 0.045293.$$

8. For physics students. The kinetic energy of a relative particle is given by $K=(\gamma-1)mc^2$ where $\gamma=\gamma=\frac{1}{\sqrt{1-\frac{v^2}{c^2}}}$. Here m is the constant mass of the Particle, v its speed and c are the constant speed of light. $v \ll c$, $k \approx 1/2mv^2$.

Solution

$$\gamma=\frac{1}{\sqrt{1-\frac{v^2}{c^2}}}$$

$$=(1-\frac{v^2}{c^2})^{-1/2}$$

$$\approx 1+\frac{1}{2}\frac{v^2}{c^2}+\frac{3}{8}(\frac{v^2}{c^2})^2+\cdots\cdots$$

$$\approx 1+\frac{1}{2}\frac{v^2}{c^2}+\frac{3}{8}(\frac{v^2}{c^2})^2+\cdots\cdots$$

and so, $k \approx (1+\frac{1}{2}\frac{v^2}{c^2}+\frac{3}{8}\frac{v^4}{c^4}+\cdots\cdots-1)\,mc^2=\frac{1}{2}mv^2+\frac{3}{8}mv^2(\frac{v}{c})^2+\cdots\cdots$

this is approximately $k \approx \frac{1}{2}mv^2$ if $v \ll c$ since the neglected terms are 'small'.

9. Find the first 3 terms in the Maclaurin series for cos (sin x). Hence or otherwise find limit $x \to 0\left(1-\frac{\cos(sinx)}{x^2}\right)$.

Solution:

From $\cos x = 1-\frac{x^2}{2!}+\frac{x^4}{4!}+\frac{x^6}{6!}+\cdots$ *and* $\sin x = x-\frac{x^3}{3!}+\frac{x^5}{5!}+\cdots$, *we have that*

$$Cos(\sin x) = 1-\frac{(x-\frac{x^3}{3!}-\frac{x^5}{5!}\cdots\cdots)^2}{2!}+(\frac{x-\frac{x^3}{3!}+\frac{x^5}{5!})^4\cdots}{4!}\cdots$$

$$=1-\frac{x^2}{2!}+\frac{2x^4}{3!2!}+\frac{x^4}{4!}+\cdot$$

$$= 1-\frac{x^2}{2}+\frac{5x^4}{24}+....$$

Thus

$$Limit_{x\to 0}\frac{1-\cos(sinx)}{x^2} = limit_{x\to 0}\frac{1-(1-\frac{x^2}{2}\frac{5x^4}{24}}{x^2}+.....$$

$$= limit_{x\to 0}\frac{\frac{x^2}{2}-\frac{5x^4}{24}}{x^2}$$

$$= limit_{x\to 0}(\frac{1}{2}-\frac{5x^2}{24})$$

$$=\frac{1}{2}$$

10. Find the values of c, which satisfies the equation $\frac{f(b)-f(a)}{b-c}$ = f'(c) as stated in mean value theorem for the function f(x)=$\sqrt{x-1}$ in the interval {1,3}

Solution

First the conditions of mean value theorem are to be checked.

f(x) is continuous in its domain [o, ∞]. and hence in the given interval [1,3]

f(x) is also differentiable in the given interval.

Plugging a=1 and b=3 in the expression on the left side of the equation,

$\frac{f(3)-f(1)}{3-1}$ = (√2-0)/2= √2/2.

Now, the derivative of the function can be found using chain rule as

$$f'(x) = \frac{1}{2\sqrt{x-1}}$$

Hence, the equation can be formed as

$$f'(c) = \frac{1}{2\sqrt{c-1}} = \sqrt{2}/2$$

cross multiplying and squaring the equation reduces to,

2(c-1) = 1 which gives the solution as c= 3/2, which lies in the given interval. [1, 3]

11. Given $f(x) = x^3$ and a = -2 and b= 2. Find the points which satisfies the mean value theorem.

Solution

$f'(x) = 3x^2$

$$f'(c) = \frac{f(b)-f(a)}{b-c}$$

$$3c^2 = \frac{8-(-8)}{2-(-2)}$$

$$3c^2 = \frac{16}{4}$$

$$3c^2 = 4$$

$$c^2 = \frac{4}{3}$$

$$c = \pm\frac{2\sqrt{3}}{3}$$

12. Find the Maclaurin series expansion of $f(x) = e^x$.

Solution

Derivative of exponential function is

$f'(x) = e^x$

in fact, all the derivatives are e^x

$f'(0) = e^0 = 1$

$f''(0) = e^0 = 1$

$f'''(0) = e^0 = 1$

we see that all the derivatives, when evaluated at x=0, gives us the value 1.

Also, f(0)=1, so we can conclude the Maclaurin series expansion will be

$$e^x \approx 1 + x + \frac{1}{2}(x^2) + \frac{1}{6}(x^3) + \frac{1}{24}(x^4) + \frac{1}{125}(x^5).$$

13. Given that y(x) is the solution to $\frac{dy}{dx} = y^3 + 2$, y(0) =3, find the value of y(0, 2) from a second order Taylor polynomial around x=0.

Solution

The second order Taylor polynomial is

$$y\,(x+h) = y(x) + y'(x)h + \frac{y''(x)}{2}(h^2)$$

x=0, h= 0.2-0 =0.2

$$y\,(0+0.2) = y\,(0) + y'(x)h + \frac{y''(0)}{2!}\,(0.2^2)$$

$$y\,(0.2) = y\,(0) + y'\,(0) \times 0.2 + y''(0) \times 0.02$$

$$y\,(0) = 3$$

$$y'(x) = y^3 + 2$$

$$y'(0) = 3^3 + 2 = 29$$

$$y''(x) = 3y^2\frac{dy}{dx} = 3y^2(y^3+2)$$

$$y''(0) = 3(3^2)\,(3^3+2) = 783$$

$$y\,(0.2) = 3+29 \times 0.2 + 783 \times 0.02$$

$$= 24.46$$

Chapter 6

Integration

Formula

1. $\int x^n \, dx = \frac{x^{n+1}}{n+1} + c, \; n \neq -1$
2. $\int dx = x + c$
3. $\int \cos x \, dx = \sin x + c$
4. $\int \sin x \, dx = -\cos x \, dx$
5. $\int \sec^2 x \, dx = \tan x + c$
6. $\int \operatorname{cosec}^2 x \, dx = -\cot x + c$
7. $\int \sec x \tan x \, dx = \sec x + c$
8. $\int \operatorname{cosec} x \cot x \, dx = -\operatorname{cosec} x + c$
9. $\int e^x dx = e^x + c$

10. $\int \frac{1}{x} dx = log|x| + c$

11. $\int tan\, x\, dx = log|sec\, x| + c \;\; or \; - log|\cos x| + c$

12. $\int cot\, x\, dx = log|sin\, x| + c$

13. $\int sec\, x\, dx = log|sec\, x + tan\, x| + c$

14. $\int cosec\, x\, dx = log|cosec\, x - cot\, x| + c$

15. $\int \frac{dx}{x^2-a^2} = \frac{1}{2a} log\left|\frac{x-a}{x+a}\right| + c$

16. $\int \frac{dx}{a^2-x^2} = \frac{1}{2a} log\left|\frac{a+x}{a-x}\right| + c$

17. $\int \frac{dx}{x^2+a^2} = \frac{1}{a} \tan^{-1}\frac{x}{a} + c$

17. $\int \frac{dx}{\sqrt{x^2-a^2}} = log\left|x + \sqrt{x^2 - a^2}\right| + c$

18. $\int \frac{dx}{\sqrt{a^2-x^2}} = \sin^{-1}\frac{x}{a} + c$

19. $\int \frac{dx}{\sqrt{x^2+a^2}} = log\left|x + \sqrt{x^2 + a^2}\right| + c$

20. $\int \sqrt{x^2 - a^2}\, dx = \frac{x}{2}\sqrt{x^2 - a^2} - \frac{a^2}{2} log\left|x + \sqrt{x^2 - a^2}\right| + c$

21. $\int \sqrt{a^2 - x^2}\, dx = \frac{1}{2} x\sqrt{a^2 - x^2} + \frac{a^2}{2} \sin^{-1}\frac{x}{a} + c$

22. $\int \sqrt{x^2 + a^2}\, dx = \frac{1}{2} x\sqrt{x^2 + a^2} + \frac{a^2}{2} log\left|x + \sqrt{a^2 + x^2}\right| + c$

INTEGRATION PROBLEMS

1. $\int(5x^2 - 8x + 5)dx = \frac{5x^3}{3} - 4x^2 + 5x + C$

2. $\int\left(x^{\frac{3}{2}} + 2x + 3\right)dx = \frac{2x^{\frac{5}{2}}}{5} + x^2 + 3x + C$

3. $\int\left(\sqrt{x} + \frac{1}{3\sqrt{x}}\right)dx = \int\left(x^{\frac{1}{2}} + \frac{1}{3x^{\frac{1}{2}}}\right)dx$

4. $\int 7\cos(5x)\,dx = \frac{7\sin(5x)}{5} + C$

5. $\int\frac{x^2}{1+x^2}dx$

Solution

$$\int\frac{x^2+1-1}{1+x^2}dx = \int\frac{x^2+1}{x^2+1}dx - \int\frac{1}{1+x^2}dx = x - \tan^{-1}x + c$$

6. $\int \sec^2 x \,.\, cosec^2 x\,dx$

Solution

$$\int sec^2 x(1 + cot^2 x)\,dx$$

$$= \int\frac{1}{cos^2 x}\left(1 + \frac{cos^2 x}{sin^2 x}\right)dx = \int\frac{1}{cos^2 x} + \int\frac{1}{sin^2 x}dx$$

$$= \int sec^2 x + \int cosec^2 x\,dx = \tan x - \cot x + c$$

7. $\int\frac{x^2-1}{x^2+1}dx$

Solution

$$\int\frac{x^2-1+1-1}{x^2+1}dx = \int\frac{x^2+1}{x^2+1}dx - \int\frac{2}{x^2+1}dx$$

Use $\left(\int\frac{dx}{x^2+1} = \tan^{-1}x + c\right)$

$$= x - \int\frac{2}{x^2+1}dx = x - 2\tan^{-1}x + c$$

8. $\int \boldsymbol{cos^4 x\, dx}$

Solution

$$\int (cos^2)^2\, x\, dx = \int \left(\frac{1+cos2x}{2}\right)^2 dx$$

$$=\frac{1}{4}\int (1+cos2x)^2\, dx = \frac{1}{4}\left[\int dx + \int cos^2 2x + \int 2cos2x\right] dx$$

$$=\frac{1}{4}\left[x + \int \frac{1+cos4x}{2}\, dx + 2\int cos2xdx\right]$$

$$=\frac{1}{4}\left[x + \frac{1}{2}\int dx + \int cos4x\, dx + 2\int cos^2 x\, dx\right]$$

$$=\frac{1}{4}\left[x + \frac{x}{2} + \frac{sin4x}{4} + \frac{2sin2x}{2}\right] + c = \frac{1}{4}\left[\frac{3x}{2} + \frac{sin4x}{4} + sin2x\right] + c$$

9. $\int \boldsymbol{sin2x\, .cos3x\, dx}$

Solution

$\mathbf{2\sin a\, .\cos b = \sin(a+b) + \sin\,(a-b)}$ **Formula**

$$= \int \frac{\sin(2x+3x)+\sin\,(2x-3x)}{2}\, dx$$

$$=\frac{1}{2}\left[\int sin5x - \int sinx\, dx\right]$$

$$=\frac{1}{2}\left[\frac{cos5x}{5} - cosx\right] + c$$

10. $\int \boldsymbol{sin3x\, .sin5x\, dx}$

Solution

We know $\mathbf{2\sin a\ .\sin b = \cos\,(a-b) - \cos\,(a+b)}$

$$= \int \frac{\cos(3x-5x)-\cos\,(3x+5x)}{2}\, dx$$

$$=\frac{1}{2}\int \cos(-2x)dx - \int cos(8x)\, dx$$

$= -\frac{1}{2}\left[\frac{sinx}{2} - \frac{sin8x}{8}\right] + c$

11. $\int \frac{\sec x}{\sec x+\tan x}\, dx$

Solution

$\int \frac{\cos x.\cos x}{\cos x(\sec x+\tan x)}\, dx \left(secx = \frac{1}{cosx},\ tanx = \frac{sinx}{cosx}\right)$

$= \int \frac{1}{1+\sin x}\, dx$

$= \int \frac{1-\sin x}{(1-\sin x)(1+\sin x)}\, dx$

$= \int \frac{1-\sin x}{cos^2 x}\, dx$

$= \int sec^2 x\, dx - \int \sec x . \tan x\ dx$

$= \tan x - \sec x + c$

12. $\int (4x^3 + 3x^2 + 2x + 4)\ dx$

Solution

$\frac{4x^4}{4} + \frac{3x^3}{3} + \frac{2x^2}{2} + 4x + c$

$= x^4 + x^3 + x^2 + 4x + c$

13. $\int \sqrt{x}(x^{3/2} - 7x^{\frac{5}{2}} + 9x^{7/2})\ dx$

Solution

$\int (x^2 - 7x^3 + 9x^4)dx$

$= \frac{x^3}{3} - \frac{7x^4}{4} + \frac{9x^5}{5} + c$

$= \frac{20x^3 - 105x^4 + 108x^5}{60} + c$

14. $\int \frac{(1+x)^3}{\sqrt{x}}\, dx \int \frac{(1+x)^3}{\sqrt{x}}\, dx$

Solution

$$\cdot\int \frac{1+x^3+3x(x+1)}{\sqrt{x}}\, dx$$

$$= \int \frac{1+x^3+3x^2+3x}{\sqrt{x}}\, dx$$

$$= \int \frac{dx}{\sqrt{x}} + \int x^{5/2}\, dx + 3\int x^{3/2}\, dx + 3\int \sqrt{x}\, dx$$

$$= 2\sqrt{x} + \frac{2x^{7/2}}{7} + \frac{6^{5/2}}{5} + 2x^{3/2} + c$$

15. $\int \frac{x^4+5x^2+3}{\sqrt{x}}\, dx$

Solution

$$\int x^{7/2}\, dx + 5\int x^{3/2}\, dx + 3\int \sqrt{x}\, dx$$

$$= \frac{2x^{9/2}}{9} + \frac{2x^5}{2} + 6\sqrt{x} + c$$

16. $\int \frac{(x^3+2^3)(x-1)}{x^2-2x+4}\, dx$

Solution

$$= \int \frac{(x+2)(x^2+4-2x)(x-1)}{(x^2-2x+4)}\, dx$$

$$= \int (x+2)(x-1)\, dx$$

$$= \int (x^2+x-2)dx = \frac{x^3}{3} + \frac{x^2}{2} - 2x + c$$

17. $\int \frac{\sin^3 x+\cos^3 x}{\sin^2 x.\cos^2 x}\, dx$

Solution

$$\frac{\sin^3 x}{\sin^2 x.\cos^2 x} - \frac{\cos^3 x}{\sin^2 x.\cos^2 x} = \frac{\sin x}{\cos x} \times \frac{1}{\cos x} - \frac{\cos x}{\sin x} \times \frac{1}{\sin x}$$

$= \tan x . \sec x - \cot x . \text{cosec}\, x$

$= \int \tan x . \sec x \, dx - \int \cot x . \text{cosec}\, x \, dx$

$= \sec x + \text{cosec}\, x + c$

18. $\int \frac{sec^2 x}{cosec^2 x} dx$

Solution

$\int \frac{1/cos^2 x}{1/sin^2 x} dx = \int \frac{sin^2 x}{cos^2 x} dx$

$= \int tan^2 x \; dx$

$= \int (sec^2 x - 1) \, dx$

$= \int sec^2 x \, dx - \int dx$

$= \tan x - x + c$

19. $\int sin^3 x \, dx$

Solution

$.\sin 3x = 3 \sin x - 4 \, sin^3 x$

$= sin^3 x = \frac{1}{4}(3 \sin x - \sin 3x)$

$= \frac{1}{4} \int (3 \sin x - \sin 3x) \; dx$

$= \frac{1}{4}\left[-3cosx + \frac{\cos 3x}{3}\right] + c$

20. $\int cos^4 x \, dx$

Solution

$.\int (cos^2 x)^2 dx$ $\quad$ $[\cos 2x = 2cos^2 x - 1]$

$$= \int \left(\frac{1+\cos 2x}{2}\right)^2 dx$$

$$= \int \frac{(1+\cos 2x)^2}{4} dx$$

$$= \int \frac{1+cos^2 x+2\cos 2x}{4} dx$$

$$= \int \frac{1+\left(1+\cos 4x/_2\right)+2\cos 2x}{4} dx$$

$$= \int \frac{2+1+\cos 4x+4\cos 2x}{8} dx$$

$$= \frac{1}{8}\left(2x + x + \frac{\sin 4x}{4} + 2\sin 2x\right) + c$$

21. $\int \sqrt{\mathbf{1} - \mathbf{\cos} \, \boldsymbol{x} \, \boldsymbol{dx}}$

Solution

$\int \sqrt{2sin^2 \frac{x}{2} \, dx}$ $[\mathbf{\cos 2}\boldsymbol{x} = \mathbf{1} - \mathbf{2}\boldsymbol{sin}^{\mathbf{2}}\boldsymbol{x} \rightarrow \mathbf{2}\boldsymbol{sin}^{\mathbf{2}}\frac{\boldsymbol{x}}{\mathbf{2}} = \mathbf{1} - \mathbf{\cos}\,\boldsymbol{x}]$

$$= \sqrt{2} \int \sin\frac{x}{2} \, dx$$

$$= -2\sqrt{2}\, cos\frac{x}{2} + c$$

INTEGRATION BY SUBSTITUTION

22. $\int (\mathbf{2}\boldsymbol{x} - \mathbf{1})^{\mathbf{5}} \boldsymbol{dx}$

Solution

Let u = 2x-1$\rightarrow$ $\frac{du}{dx} = 2$ make dx the subject

$$dx = \frac{du}{2}$$

$replace\ 2x - 1\ by\ u$ and write dx in terms of du

$\int (2x-1)^5\, dx = \int u^5 \frac{du}{2}$

$= \frac{1}{2} \int u^5\, du$

$= \frac{1}{2} \left[\frac{u^6}{6}\right] + c$

$= \frac{1}{12} u^6 + c$

$= \frac{1}{12} (2x-1)^6 + c$

23. $\int \frac{1}{x+\sqrt{x}} dx$

Solution

Substitute $X=u^2$

$$\frac{dx}{du} = 2u$$

$$= \int \frac{1}{u^2 + \sqrt{u^2}} 2udu$$

$$= \int \frac{1}{u^2 + u} 2udu$$

$$= \int \frac{2}{u+1} du$$

$= 2ln(u+1) + c$

$I = 2ln(\sqrt{x} + 1) + c$

24. $\int e^{-x^2} x\, dx$

Solution

$u=x^2$ then 2x dx= du

$$\int e^{-x^2} x\,dx = \int e^{-u} \frac{du}{2}$$

$$= \frac{1}{2}\int e^{-u} du$$

$$= \frac{1}{2}(-e^{-u}) + c$$

$$= -\frac{1}{2}e^{-u} + c$$

$$= -\frac{1}{2}e^{-x^2} + c$$

25 $\int 2x\sqrt{1+x^2}dx$

Solution

u=1+x^2

$2xdx = du$

$$\int 2x\sqrt{1+x^2}dx = \int \sqrt{u}du$$

$$= \int u^{\frac{1}{2}} du$$

$$= \frac{u^{\frac{3}{2}}}{\frac{3}{2}} + c$$

$$= \frac{2}{3} u^{\frac{3}{2}} + c$$

$$= \frac{2}{3}(1+x^2) + c$$

26. $\int x(a-x)^8 dx$

Solution

$u=a-x \qquad x=a-u$

$du=-dx$

$$\int x(a-x)^8dx = \int (a-u)u^8/(-du)$$

$$= \int(-a(u)^8 + u^9)du$$

$$= \int u^9du - a\int u^8\, du$$

$$= \frac{u^{10}}{10} - a\frac{u^9}{9} + c$$

$$= \frac{(a-x)^{10}}{10} - \frac{a(a-x)^9}{9} + c$$

27. $\int sinxcosx\, dx$

Solution

Let, u=sin x

$$\frac{du}{dx} = cosx$$

du=cos x dx

$$\int \sin x \cos x\, dx = \int udu$$

$$= \frac{u^2}{2} + c$$

$$= \frac{\sin x^2}{2} + c$$

28. $\int \frac{e^x}{e^x-1} dx$

Solution

$u = e^x - 1$

$$\frac{du}{dx} = e^x$$

du=$e^x\, dx$

$$\int \frac{e^x}{e^x - 1} dx = \int \frac{du}{u}$$

$$= \log|u| + c$$

$$= \log|e^x - 1| + c$$

29. $\int \frac{1}{x \log x} dx$

Solution

Let, $\log x = u$

$$\frac{1}{x} dx = du$$

$$\int \frac{1}{x \log x} dx = \int \frac{du}{u}$$

$$= \log |u| + c$$

$$= \log |\log x| + c$$

30. $\int \tan x \sqrt{\sec x}\, dx$

Solution

$$u = \sec x$$

$$\frac{du}{dx} = \sec x \tan x$$

$$dx = \frac{du}{\sec x \tan x}$$

$$\int \tan x \sqrt{\sec x}\, dx = \int \tan x \sqrt{u} \frac{du}{\sec x \tan x}$$

$$= \int \frac{\sqrt{u}}{u} du$$

$$= \int \frac{\sqrt{u}}{\sqrt{u}\sqrt{u}} du$$

$$= \int \frac{1}{\sqrt{u}} du$$

$$= \int u^{\frac{-1}{2}} du$$

$$= \frac{u^{\frac{-1}{2}+1}}{\frac{-1}{2}+1} + c$$

$$= \frac{u^{\frac{1}{2}}}{\frac{1}{2}} + c$$

$$= \frac{1}{2} u^{\frac{1}{2}} + c$$

$$= \frac{1}{2}\sqrt{u} + c$$

=

31. $\int \frac{1}{x(logx)log\,(logx)} dx$

Solution

Let, $\mathbf{u} = log(logx)$

$$\frac{du}{dx} = \frac{1}{logx}\ x\ \frac{1}{x}$$

$$dx = x\, logx\, du$$

$$\int \frac{1}{x(logx)log\,(logx)} dx = \int \frac{du}{u}$$

$$= log\,|u| + c$$

$$= log|log(logx)| + c$$

32. $\int (5x+3)\sqrt{(2x-1)}\ dx$

Solution

Let $z = 2x - 1;\ x = z + \frac{1}{2} \Rightarrow \frac{dz}{dx} = 2 \Rightarrow dx = \frac{dz}{2}$

$= \int \left[\frac{5(z+1)}{2} + 3\right]\sqrt{z}\ dz$

$= \int \left(\frac{5z+5}{2} + 3\right)\sqrt{z}\ dz = \int \left(\frac{5z+11}{2}\right)\sqrt{z}\ dz$

$= \frac{5}{2}\int \sqrt{z}\, dz + \frac{11}{2}\int \sqrt{z}\ dz$

$= \frac{5}{3} z^{\frac{3}{2}} + \frac{11}{3} z^{\frac{3}{2}} + c$

$= \frac{5}{3}(2x-1)^{\frac{3}{2}} + \frac{11}{3}(2x-1)^{\frac{3}{2}} + c$

33. $\int \frac{x^2}{\sqrt{x+2}}\ dx$

Solution

Let $z = x + 2 \Rightarrow x = z - 2 \Rightarrow \frac{dz}{dx} = 1$

$= \int \frac{(z+2)^2}{\sqrt{z}}\ dz$

$= \int \frac{z^2+4+4z}{\sqrt{z}}\ dz$

$= \int z^{\frac{3}{2}}\ dz + 4\int z^{\frac{-1}{2}}\ dz + 4\int \sqrt{z}\ dz$

$= \frac{2z^{\frac{5}{2}}}{5} + 8\sqrt{z} + \frac{8z^{\frac{3}{2}}}{3} + c$

$= \frac{2(x+2)^{\frac{5}{2}}}{5} + 8\sqrt{x+2} + \frac{8(x+2)^{\frac{3}{2}}}{3} + c$

34. $\int (x+2)\sqrt{3x+5}\,dx$

Solution

Let $z = 3x + 5 \Rightarrow x = \frac{z-5}{3} \Rightarrow \frac{dz}{dx} = 3 \Rightarrow dx = 3dz$

$$= \int \left(\frac{z-5}{2} + 2\right)\sqrt{z}\, 3dz$$

$$= \int \left(\frac{z-5+6}{3}\right)\sqrt{z}\, 3dz$$

$$= \int \left(\frac{z+1}{3}\right)\sqrt{z}\, 3dz$$

$$= \int \frac{z^{\frac{3}{2}}}{3} + \frac{z^{\frac{1}{2}}}{3}\ 3dz$$

$$= \frac{1}{3} \times \frac{z^{\frac{5}{2}}}{\frac{5}{2}} + \frac{z^{\frac{3}{2}}}{\frac{2}{3}} + c$$

$$= \frac{2z^{\frac{5}{2}}}{15} + \frac{2z^{\frac{3}{2}}}{3} + c$$

$$= \frac{2}{15}(3x+5)^{\frac{5}{2}} + \frac{2}{3}(3x+5)^{\frac{3}{2}} + c$$

35. $\int (x-1)\sqrt{x+1}\ dx$

Solution

Let $z = x + 1 \Rightarrow \frac{dz}{dx} = 1 \Rightarrow x = z - 1$

$$= \int (z-1-1)\sqrt{z}\ dz$$

$$= \int (z-2)\sqrt{z}\, dz$$

$$= \int z^{\frac{3}{2}}\, dz - \int 2\sqrt{z}\, dz$$

$$= \frac{z^{\frac{5}{2}}}{\frac{5}{2}} - \frac{2z^{3/2}}{\frac{3}{2}} + c$$

$$= \frac{2(x+1)^{5/2}}{5} - \frac{4(x+1)^{3/2}}{3} + c$$

36. $\int \frac{x}{\sqrt{x+5}}\, dx$

Solution

Let $z = x + 5 \Rightarrow x = z - 5 \Rightarrow \frac{dz}{dx} = 1 \Rightarrow dx = dz$

$$= \int \frac{z-5}{\sqrt{z}}\, dz$$

$$= \int \sqrt{z}\, dz - 5 \int z^{-1/2}\, dz$$

$$= \frac{z^{3/2}}{3/2} - \frac{5z^{1/2}}{1/2} + c$$

$$= \frac{2}{3}(x+5)^{\frac{3}{2}} - 10(x+5)^{\frac{1}{2}} + c$$

37. $\int \frac{x}{\sqrt{1-2x}}\, dx$

Solution

Let $z = 1 - 2x \Rightarrow x = \frac{1-z}{2} \Rightarrow \frac{dz}{dx} = -2 \Rightarrow dx = \frac{-dz}{2}$

$$= \int \frac{1-z/2}{\sqrt{z}} \times \frac{-1}{2}\, dz$$

$$= -\int \frac{1-z}{4\sqrt{z}}\, dz$$

$$= -\frac{1}{4}\left[\int z^{-1/2}\, dz - \int \sqrt{z}\, dz\right]$$

$$= -\frac{1}{4}\left[\frac{\sqrt{z}}{1/2} - \frac{z^{3/2}}{3/2}\right] + c$$

$$= -\frac{1}{4}\left[2\sqrt{z} - \frac{2z^{\frac{3}{2}}}{3}\right] + c$$

$$= \frac{1z^{\frac{3}{2}}}{6} - \frac{1}{2}\sqrt{z} + c$$

$$= \frac{1}{6}(1-2x)^{\frac{3}{2}} - \frac{1}{2}(1-2x)^{\frac{1}{2}} + c$$

SUBSTITUTION RULES FOR SINE, COSINE, TAN AND COT

Rules for $\int sin^m x\ sin^n x\ dx$

Rule 1: If power of sinx is odd and positive integer, then take z=cos x

Rule 2: If power of cosx is odd and positive integer, then take z=sin x

Rule 3: If power of both sinx & cosx is odd and positive integer, then take z= sin x or cosx

Rule 4: If power of neither sinx or cosx is odd and positive integer, then see the sum of power of both sin & cos, if even negative came then put z=tan x

Rule 5. If sum of powers of both sinx and cosx is even positive integers then express the integrand as sum of sines and cosines of multiple angles using the following formula

$$sin^2 x = \frac{1-cos2x}{2},\quad cos^2 x = \frac{1+cos2x}{2},$$

$$sin^3 x = \frac{3sinx - sin3x}{4},\quad cos^3 x = \frac{3cosx + cos3x}{4}$$

1. $\int sin^5 x\, dx$

Solution

$$.\int sin^5 x . \cos x\ dx$$

According to rule 1, $z = \cos x \rightarrow \frac{dz}{dx} = -\sin x \rightarrow dx = \frac{dz}{-\sin x}$

$$= -\int sin^5 x \frac{dz}{sin\ x}$$

$$= -\int sin^4 x\, dz$$

$$= -\int (1 - z^2)^2\, dz$$

$$= -\int (1 + z^4 + 2z^2)\, dz$$

$$= -\left[z + \frac{z^5}{5} - 2\frac{z^3}{3}\right] + c$$

$$= \frac{2cos^3 x}{3} - \frac{cos^5 x}{5} - \cos x - c$$

2. $\int cos^3 x . \sqrt{\sin x}\ dx$

Solution

According to rule 2, $z = \sin x \rightarrow \frac{dz}{dx} = \cos x \rightarrow dx = \frac{dz}{\cos x}$

$$= \int cos^3 x \sqrt{\sin x}\ dx$$

$$= \int (\cos x)(1 - sin^2 x)\sqrt{\sin x}\ dx \quad \mathbf{cos^2 x = 1 - sin^2 x}$$

$$= \int \cos x \left(\sqrt{sinx} - sin^{\frac{5}{2}} x\right) dx$$

$$= \int \cos x \left(\sqrt{z} - z^{\frac{5}{2}}\right) \frac{dz}{\cos x}$$

$$= \int \left(\sqrt{z} - z^{\frac{5}{2}}\right) dz$$

$$= \frac{2z^{\frac{3}{2}}}{3} - \frac{2z^{\frac{7}{2}}}{7} + c$$

$$= \frac{2sin^{\frac{3}{2}} x}{3} - \frac{2sin^{\frac{7}{2}}}{7} + c$$

3. $\int sin^3x.cos^2x\,dx$

Solution

According to rule 1, $z = \cos x \rightarrow \frac{dz}{dx} = -\sin x \rightarrow dx = \frac{-dz}{\sin x}$

$= -\int sin^3x.cos^2x\frac{dz}{\sin x}$

$= \int z^2\,dz - \int z^4\,dz$

$= \frac{z^3}{3} - \frac{z^5}{5} + c$

$= \frac{cos^3 x}{3} - \frac{cos^5 x}{5} + c$

4. $\int sin^3x.cos^5x\,dx$

Solution

$.\int \sin x(1 - cos^2x)cos^5x\,dx$ $[sin^2x = 1 - cos^2x]$

$= \int \sin x(cos^5x - cos^7x)\,dx$

According to rule 1, $z = \cos x \rightarrow \frac{dz}{dx} = -\sin x \rightarrow dx = \frac{-dz}{\sin x}$

$= -\int \sin x(z^5 - z^7)\frac{dz}{\sin x}$

$= -\int (z^5 - z^7)\,dz$

$= -\frac{z^6}{6} + \frac{z^8}{8} + c$

$= \frac{cos^8 x}{8} - \frac{cos^6 x}{6} + c$

5. $\int sin^4 x\, dx$

Solution

.Reduction formula:

$$\int sin^n x\, dx = \frac{n-1}{n}\int[sin^{n-2}x\, dx] - \frac{\cos x\, sin^{n-1}x}{n}$$

= Here, n=2

$= -\frac{\cos x\ sin^3 x}{4} + \frac{3}{4}\int sin^2 x\, dx$

= for $\int sin^2 x\, dx$ apply reduction formula again,

$= -\frac{\cos x \sin x}{2} + \frac{1}{2}\int dx$

$= \frac{x}{2} - \frac{\cos x \sin x}{2}$

$= \int sin^4 x\ dx = -\frac{\cos x\ sin^3 x}{4} + \frac{3}{4}\int sin^2 x\ dx$

$= -\frac{\cos x\ sin^3 x}{4} - \frac{3\cos x \sin x}{8} + \frac{3x}{8} + c$

$= \frac{\sin 4x - 8\sin 2x + 12x}{32} + c$

(This problem can also be solved by using the Rule 5 of $\int \sin^m x.\cos^n x dx$)

6. $\int sin^3 x . \sin 2x\, dx$

Solution

$.\int 2sin^4 x . \cos x\, dx$ [$\sin 2x = 2\sin x . \cos x$]

According to rule 2, $z = \sin x \rightarrow \frac{dz}{dx} = \cos x \rightarrow dx = \frac{dz}{\cos x}$

$= \int 2z^4 \cos x\ \frac{dz}{\cos x}$

$= \int 2z^4\, dz$

$$= \frac{2z^5}{5} + c$$

$$= \frac{2sin^5 x}{5} + c$$

7. $\int cos^2x\,.\sin 3x\,dx$

Solution

As $\sin 3x = 3cos^2x.\sin x - sin^3x$

$$= \int cos^2x(3cos^2x.\sin x - sin^3x)\,dx$$

$$= \int cos^2x(3cos^2x.\sin x - \sin x + \sin x.cos^2x)\,dx$$

$$= \int (3cos^4x.\sin x - cos^2x\,.\sin x + cos^4x.\sin x)\,dx$$

$$= \int \sin x(4sin^4x - cos^2x)\,dx$$

According to rule 1, $z = \cos x \rightarrow \frac{dz}{dx} = -\sin x \rightarrow dx = \frac{-dz}{\sin x}$

$$= -\int \sin x(4z^4 - z^2)\frac{dz}{\sin x}$$

$$= -\int (4z^4 - z^2)\,dz$$

$$= -\left[\frac{4z^5}{5} - \frac{z^3}{3}\right] + c = \frac{cos^3 x}{3} - 4\frac{coc^5 x}{5} + c$$

Rules for $\int \tan^m x.\sec^n x dx$

Rule 1: If power of sec x is even positive integer, then z= tan x

Rule 2: If power of sec x is not even positive integer, then see the power of tan:

- ☆ If power of tan x is odd positive integer, then z= sec x
- ☆ If power of tan x is even positive integer, then use the formula $tan^2x = sec^2x - 1$ and then take, z= tan x

Rule 3: If power of tanx is zero and secx is odd positive integer greater then 1 use integration by parts methods

8. $\int \boldsymbol{tan^2x.sec^4\ dx}$

Solution

Acc to rule 1, $z = \tan x \to \frac{dz}{dx} = sec^2x \to dz = sec^2x\,dx$

$= \int tan^2x\,(1 + tan^2x)sec^2x\,dx$

We know $[\boldsymbol{tan^2x + 1 = sec^2x}]$

$= \int z^2(1 + z^2)sec^2x\,dx$

$= \int z^2(1 + z^2)\,dz$

$= \frac{z^3}{3} + \frac{z^5}{5} + c$

$= \frac{tan^3x}{3} + \frac{tan^5x}{5} + c$

9. $\int \boldsymbol{tan^3x.sec^2x\ dx}$

Solution

According to rule 1, $z = \tan x \to \frac{dz}{dx} = sec^2x \to dz = sec^2x\,dx$

$= \int sec^2x(1 + tan^2x)\ dx$

$= \int (1 + z^2)\,dz$

$= z + \frac{z^3}{3} + c$

$= \tan x + \frac{tan^3x}{3} + c$

10. $\int \boldsymbol{tan^3x\ dx}$

Solution

$.\int tan^2x.\tan x\,dx$

$= \int (sec^2x - 1)\tan x\ dx$

$$= \int \tan x . sec^2 x\, dx - \int tan\, x\, dx \qquad (1)$$

$For, \int \tan x . sec^2 x\, dx$:

According to Rule 1, $z = \tan x \rightarrow \frac{dz}{dx} = sec^2 x$

From (1)

$$= \int z\, dz - log|\sec x| + c \; [\int tan\, x\, dx = log|sec\, x| + c]$$

$$= \frac{z^2}{2} - log|\sec x| + c$$

$$= \frac{tan^2 x}{2} - \log|\sec x| + c$$

11. $\int tan^4 x\, dx$

Solution

$$.\int (sec^2 x - 1)\, tan^2 x\, dx$$

According to rule 1, $z = \tan x \rightarrow \frac{dz}{dx} = sec^2 x \rightarrow dx = \frac{dz}{sec^2 x}$

$$= \int sec^2 x\, tan^2 x\, dx - \int tan^2 x\, dx$$

$$= \int sec^2 x\, tan^2 x \frac{dz}{sec^2 x} - \int (sec^2 x - 1)\, dx$$

$$= \int tan^2 x\; dz - \int sec^2 x\, dx + \int dx$$

$$= \int z^2\; dz - \int sec^2 x\, dx + \int dx$$

$$= \frac{tan^3 x}{3} - \tan x + x + c$$

11. $\int tan^4 x . sec^2 x\, dx$

Solution

According to rule 1,

$$z = \tan x \rightarrow \frac{dz}{dx} = sec^2 x\, dx \rightarrow dz = sec^2 x\, dx$$

$= \int z^4 dz$

$= \frac{z^5}{5} + c$

$= \frac{tan^5 x}{5} + c$

Rules for $\int cosec^m x . \cot^n x dx$

Rule 1: If power of cosec x is even positive integer, then, z = cot x

Rule 2: If power of cosec x is not even positive integer, then see the power of cot x. If power of cot x is odd then, z = cosec x. If power of cot x is even then, z = cot x

Rule 3: If power of cotx is zero and cosecx is odd positive integer greater than 1 then use integration by parts methods

12. $\int cosec^4 x \, dx$

Solution

$.\int (cot^2 x + 1)\, cosec^2 x \, dx$ [$cosec^2 x = cot^2 x + 1$]

According to rule 1,

$z = \cot x \rightarrow \frac{dz}{dx} = -csc^2 x \rightarrow dz = -cosec^2 x \, dx$

$= \int (-z^2 - 1)\, dz$

$= -\int z^2 \, dz - \int dz$

$= -\frac{z^3}{3} - z + c$

$= -\frac{cot^3 x}{3} - \cot x + c$

13. $\int cot^2 x . cosec^4 x \, dx$

Solution

$.\int cot^2 x\,(cot^2 x + 1) cosec^2 x \, dx$

According to rule 1,

$$z = \cot x \to \frac{dz}{dx} = -cosec^2 x \to dz = -cosec^2 x\, dx$$

$$= -\int z^2 (z^2+1)\, dz$$

$$= -\int (z^4 + z^2) dz$$

$$= -\frac{z^5}{5} - \frac{z^3}{3} + c - \frac{z^5}{5} - \frac{z^3}{3} + c$$

$$= -\frac{cot^5 x}{5} - \frac{cot^3 x}{3} + c - \frac{cot^5 x}{5} - \frac{cot^3 x}{3} + c$$

14. $\int cot^3 x\, dx$

Solution

$$\int \cot x\,(csc^2 x - 1)\, dx$$

According to rule 2,

$$z = \operatorname{cosec} x \to \frac{dz}{dx} = -\cot x.\operatorname{cosec} x \to dx = -\frac{dz}{\cot x.\operatorname{cosec} x}$$

$$= -\int \cot x\,(csc^2 x - 1)\frac{dz}{\cot x.\operatorname{cosec} x}$$

$$= -\int \frac{csc^2 x - 1}{\operatorname{cosec} x}\, dz$$

$$= -\int \frac{z^2 - 1}{z}\, dz$$

$$= -\int z\, dz + \int \frac{dz}{z}$$

$$= \log z - \frac{z^2}{2} + c$$

$$= \log|\operatorname{cosec} x| - \frac{csc^2 x}{2} + c$$

15. $\int \tan x \, dx$

Solution

$.\int \sin x . \cos^{-1} x \, dx$

According to rule 1,

$$z = \cos x \rightarrow \frac{dz}{dx} = -\sin x \rightarrow dz = -\sin x \, dx$$

$= -\int \frac{dz}{z}$

$= -\log z + c$

$= -log|\cos x| + c$

16. $\int \cot x \, dx$

Solution

$.\int \frac{\sin x}{\cos x} dx$

$$z = \sin x \rightarrow \frac{dz}{dx} = \cos x \rightarrow dz = \cos x \, dx$$

$= \int \frac{1}{z} dz$

$= \log z + c$

$= log|\sin x| + c$

17. $\int \sec x \, dx$

Solution

$.\int \frac{\sec x(\sec x + \tan x)}{\sec x + \tan x}$

$$z = \sec x + \tan x \rightarrow \frac{dz}{dx} = \sec x . \tan x + sec^2 x \rightarrow dz$$

$$= \sec x(\sec x + \tan x) \, dx$$

$= \int \frac{dz}{z}$

$= \log z + c$

$= log|\sec x + \tan x| + c$

18. $\int \mathbf{cosec}\, x\, dx$

Solution

$\cdot \int \frac{\text{cosec } x(\text{cosec } x - \cot x)}{\text{cosec } x - \cot x}$

$z = \text{cosec} x - \cot x \rightarrow \frac{dz}{dx} = -\text{cosec} x \cot x + csc^2 x \rightarrow dz$

$= \text{cosec} x(\text{cosec} x - \cot x)\, dx$

$= \int \frac{dz}{z}$

$= \log z + c$

$= log|\text{cosec} x - \cot x| + c$

19. $\int \sqrt{1 - x^2}\, dx$

Solution

$\textbf{Formula}: \int \sqrt{a^2 - x^2}\, dx = \frac{x}{2}\sqrt{a^2 - x^2} + \frac{a^2}{2} \sin^{-1}\left(\frac{x}{a}\right) + c$

$Here, a = 1$

$= \frac{x}{2}\sqrt{1 - x^2} + \frac{1}{2} \sin^{-1} x + c = \frac{x\sqrt{1-x^2} + \sin^{-1} x}{2} + c$

20. $\int \frac{dx}{\sqrt{1-x^2}}$

Solution

$$\textbf{Formula: } \int \frac{dx}{\sqrt{a^2-x^2}} = \sin^{-1}\left(\frac{x}{a}\right) + c$$

$= Here: \sin^{-1} x + c$

21. $\int \frac{dx}{\sqrt{4-x^2}}$

Solution

$Here, a = 2: \sin^{-1} \frac{x}{2} + c$

22. $\int \frac{dx}{9+x^2}$

Solution

$$\textbf{Formula: } \int \frac{dx}{a^2+x^2} = \frac{1}{a}\tan^{-1}\left(\frac{x}{a}\right) + c$$

$= Here, a = 3: \frac{1}{3}\tan^{-1}\left(\frac{x}{3}\right) + c$

23. $\int \frac{dx}{\sqrt{1-2x-x^2}}$

Solution

$\int \frac{dx}{\sqrt{1-(x^2+2x)}}$

$= \int \frac{dx}{\sqrt{1-(x^2+2x+1)+1}}$

$= \int \frac{dx}{\sqrt{2-(x+1)^2}}$

$= Here, a = \sqrt{2} \,\&\, X = x + 1: \sin^{-1}\left(\frac{x+1}{\sqrt{2}}\right) + c$

24. $\int \sqrt{2x^2 + 3x + 4}\, dx$

Solution

$$\int \sqrt{2\left(x^2 + \frac{3x}{2} + 2\right)}\, dx$$

$$= \int \sqrt{2\left[x^2 + 2 \times \frac{3}{4} \times x + \left(\frac{3}{4}\right)^2 - \left(\frac{3}{4}\right)^2 + 2\right]}\, dx$$

$$= \sqrt{2} \int \sqrt{\left(x + \frac{3}{4}\right)^2 - \frac{23}{16}}\, dx$$

$$= Here, X = x + \frac{3}{4} \;\&\; a = \frac{\sqrt{23}}{4}$$

Formula: $\int \sqrt{x^2 - a^2}\, dx = \frac{x}{2}\sqrt{x^2 - a^2} - \frac{a^2}{2} \log\left(x + \sqrt{x^2 - a^2}\right) + c$

$$= \sqrt{2}\left[\frac{4x+3}{8}\sqrt{\left(x + {}^{3}/_{4}\right)^2 - \left(\frac{23}{16}\right)} - \frac{23}{32}\right.$$

$$\left.\log\left\{x + \frac{3}{4}\sqrt{\left(x + {}^{3}/_{4}\right)^2 - \left(\frac{23}{16}\right)}\right\}\right] + c$$

25. $\int \frac{dx}{x^2 + 5x + 6}$

Solution

$$\int \frac{dx}{x^2 + 2 \times \frac{5}{2} \times x + \left(\frac{5}{2}\right)^2 - \left(\frac{5}{2}\right)^2 + 6}$$

$$= \int \frac{dx}{\left(x + \frac{5}{2}\right)^2 - \frac{1}{4}}$$

Formula: $\frac{1}{2a} \log\left(\frac{x-a}{x+a}\right)$

$$Here, X = x + \frac{5}{2} \& a = \frac{1}{2}$$

$$= log\left[\frac{(x+{}^{5}/_{2})+\frac{1}{2}}{(x+{}^{x}/_{2})+\frac{1}{2}}\right] + c$$

26. $\int \frac{dx}{2x^2+3x+4}$

Solution

$$\int \frac{dx}{2\left[x^2+2\times\frac{3}{4}\times x+\left(\frac{3}{4}\right)^2-\left(\frac{3}{4}\right)^2+2\right]}$$

$$= \frac{1}{2}\int \frac{dx}{\left(x+\frac{3}{4}\right)^2+\frac{23}{16}}$$

$$\textbf{Formula}: \frac{1}{a}tan^{-1}\left(\frac{x}{a}\right) + c$$

$$Here, X = x + \frac{3}{4} \& a = \frac{\sqrt{23}}{4}$$

$$= Acc\ to\ formula: \frac{4}{\sqrt{23}}\tan^{-1}\left(\frac{x+\frac{3}{4}}{\sqrt{23}/_{4}}\right) + c$$

INTEGRATION BY PARTS

$$\int uvdx = u\int vdx \int\left[\left(\frac{du}{dx}\right)\int vdx\right]dx$$

["u" and "v" are different function]

ILATE Rule

I = Inverse
L = Logarithmic differentiation
A = Algebraic
T = Trigonometric
E = Exponential

If both are trigonometric, take U which comes first in the the word ILATE take those function as v whose integrals is simpler.

If both are algebraic, take function as u whose differentiation is simpler

1. $\int xe^x\,dx$ **Here x as algebraic and exp(x) is exponential, x can be taken as u**

Solution

$$x\int e^x dx - \int\left[\frac{d}{dx}x \times \int e^x dx\right]dx$$

$$= x\int e^x dx - \int\left[1 \times \int e^x dx\right]dx$$

$$= x\,e^x - \int e^x dx$$

$$= x\,e^x - e^x + C$$

$$= e^x(x-1) + C$$

2. $\int \sin^{-1} x\,dx$

Solution

In case of inverse function, 1 can be taken as second function (v) and Integration by parts method can be used

$\boldsymbol{here, u} = sin^{-1}x$, *v=1, use the below method*

$$\int \boldsymbol{uvdx} = \boldsymbol{u}\int \boldsymbol{vdx} - \int\left[\left(\frac{\boldsymbol{du}}{\boldsymbol{dx}}\right)\int \boldsymbol{vdx}\right]\boldsymbol{dx}$$

$$\int sin^{-1}x\,dx = x\,sin^{-1}x - \int \frac{x}{\sqrt{1-x^2}}dx$$

$$t = 1-x^2$$

$$\frac{dt}{dx} = -2x$$

$$dt = -2xdx$$

$$xdx = -\frac{dt}{2}$$

$$\int sin^{-1}x\,dx = x\,sin^{-1}x - \int \frac{x}{\sqrt{1-x^2}}dx$$

$$= x\sin^{-1}x - \int \frac{-dt}{2\sqrt{t}}$$

$$= x\sin^{-1}x + \frac{1}{2}\int \frac{dt}{\sqrt{t}}$$

$$= x\sin^{-1}x + \frac{1}{2}\int t^{\frac{-1}{2}}dt$$

$$= x\sin^{-1}x + \frac{1}{2}\left[\frac{t^{\frac{-1}{2}+1}}{\frac{-1}{2}+1}\right] + c$$

$$= x\sin^{-1}x + \frac{1}{2}\left[\frac{t^{\frac{1}{2}}}{\frac{1}{2}}\right] + c$$

$$= x\sin^{-1}x + t^{\frac{1}{2}} + c$$

$$= x\sin^{-1}x + \sqrt{t} + c$$

$$= x\sin^{-1}x + \sqrt{1-x^2} + c$$

3. $\int xe^{-6x}dx$

Solution

Here u = "x" and v = exp(-6x), so use Integration by parts method

$$\int xe^{-6x}dx = -\left(\frac{1}{6}\right)xe^{-6x} - \int -\frac{e^{-6x}}{6}dx$$

$$= -\left(\frac{1}{6}xe^{-6x}\right) + \frac{1}{6}\int e^{-6x}dx$$

$$= -\frac{1}{6}xe^{-6x} + \frac{1}{6}\left(-\frac{e^{-6x}}{6}\right) + c$$

$$= -\frac{1}{6}xe^{-6x} - \frac{e^{-6x}}{36} + c$$

4. $\int x \sin 3x\, dx$

Solution

Let, u=x, v =sin3x and use Integration by parts methods

$$\int x\sin 3x\, dx = \frac{-x\cos 3x}{3} - \int \frac{-\cos 3x}{3}dx$$

$$= \frac{-x\cos 3x}{3} + \int \frac{\cos 3x}{3} dx$$

$$= \frac{-x\cos 3x}{3} + \frac{1}{3}\int \cos 3x dx$$

$$= \frac{-x\cos 3x}{3} + \frac{1}{3}\left[\frac{\sin 3x}{3}\right] + c$$

$$= \frac{-x\cos 3x}{3} + \frac{1}{9}[\sin 3x] + c$$

5. $\int 2x^3 e^{x^2} dx$

Solution

$$\int 2x^3 e^{x^2} dx = \int x^2 e^{x^2} . 2x\, dx$$

Let, $t = x^2$

$$\frac{dt}{dx} = 2x$$

$dt = 2xdx$

$$\int x^2 e^{x^2} 2x\, dx = \int te^t dt$$

Here $u = t$ and $v = e^t$, use Integration by parts method

$$\int te^t dt = te^t - \int\left(\frac{dt}{dt}\right)\int e^t dt$$

$$= te^t - e^t + c$$

$$= x^2 e^{x^2} - e^{x^2} + c$$

6. $\int x \cos x\, dx$

Solution

$$\int x\cos x\, dx = x\int \cos x\, dx - \int [(\frac{dx}{dx}) \int \cos x\, dx]\, dx$$

$$= x\sin x - \int \sin x dx = x\sin x + \cos x + C$$

7. $\int \log x . dx$

Solution

$$\int (\log x .1) = \log x \int 1\, dx - \int \left(\frac{d}{dx}\left(\log x\right) \int 1\, dx \right) dx$$

$$=(\log x)x - \int \frac{1}{x} x\, dx = x \log x - x + c$$

8. $\int e^x [\tan^{-1} x + (1/(1 + x^2)]\, dx$

Solution

$$I = \int e^x\, [\tan^{-1} x + 1/(1 + x^2)\,]dx$$

$$let\ f(x) = \tan^{-1} x\ , f'(x) = 1/(1 + x^2)$$

Therefore $\int e^x [(f(x) + f'(x)]dx = e^x \tan^{-1} x + C$

9. $\int \frac{(x^2 + 1)e^x}{(x + 1)^2} dx$

Solution

$$I = \int \frac{(x^2 + 1)e^x}{(x + 1)^2} dx$$

$$= \int e^x [(x^2 - 1 + 1 + 1)/(x + 1)^2\,]\,\mathrm{dx}$$

$$= \int e^x [(x^2 - 1)/(x + 1)^2 + 2/(x + 1)^2\,]dx$$

$$= \int e^x [\{(x + 1)(x - 1))/(x + 1)^2\} + 2/(x + 1)^2\,]dx]$$

$$= \int e^x [\{(x - 1))/(x + 1)\} + 2/(x + 1)^2\,]dx]$$

Let f(x)= (x-1)//(x+1), $f'(x)$=2/$(x+1)^2$

Using the formula

$$\int e^x[(f(x)+f'(x)]dx = e^x f(x)+c$$

I =[(x-1)/(x+1)] $e^x + c$

10. $\int x\sec^2 x.dx = x\int \sec^2 x.dx - \int\left[\left(\frac{dx}{dx}\right)\int [\sec^2 x\ dx]dx\right]$

Solution

$= x(\tan x) - \int 1.\tan x\,dx$

$= x\tan x - \int \tan x\,dx$

$= x\tan x + \log|\cos x| + c$

11. $\int x\sin 3x dx$

Solution

$x\int \sin 3x dx - \int\left[\left(\frac{dx}{dx}\right)\int [\sin 3x.dx]dx\right]$

$= x * \left(\frac{-\cos 3x}{3}\right) + \left[\int \frac{1}{3}\cos 3x\right]dx$

$= -\frac{x}{3}\cos 3x + \frac{1}{3}\sin 3x + C$

12. $\int x\cos^2 x dx$

Solution

We know $\int \cos^2 x dx = \frac{1+\cos 2x}{2}$

$$I = \int x\cos^2 xdx = \int x\left(\frac{1+\cos 2x}{2}\right)dx = \int\left[\frac{x}{2}+\frac{x}{2}\cos[2x]\right]dx$$

$$= \frac{1}{2}\int xdx + \frac{1}{2}\int x\cos 2xdx$$

Use the formula $\int uvdx = u\int vdx - \int\left[\left(\frac{du}{dx}\right)\int vdx\right]dx$

$$\frac{1}{2}*\frac{x^2}{2}+\frac{1}{2}\left[x\int\cos 2xdx - \int\left\{\frac{d}{dx}(x)\int[\cos 2xdx\right\}dx\right]$$

$$\frac{1}{2}*\frac{x^2}{2}+\frac{1}{2}\left[x*\left(\frac{\sin 2x}{2}\right)-\int\frac{\sin 2x}{2}dx\right]$$

$$\frac{x^2}{4}+\frac{1}{2}\left[x*\left(\frac{\sin 2x}{2}\right)-\frac{1}{2}*\frac{(-\cos[2x))}{2}\right]+c$$

$$\frac{x^2}{4}+\frac{1}{2}\left[x*\frac{\sin 2x}{2}+\frac{\cos 2x}{4}\right]+C$$

$$\frac{x^2}{4}+x*\frac{\sin 2x}{4}+\frac{\cos 2x}{8}+C$$

13. $\int x\sin 2x\cos 3xdx$

Solution

$$As\ \int uvdx = u\int vdx - \int[(\frac{du}{dx})\int vdx]dx$$

$$I = x\int\sin 2x\cos 3x.dx - \int\left[\left(\frac{dx}{dx}\right)\int[\sin 2x.\cos 3x.dx]\right]dx \quad ....(1)$$

Now, $\int\sin 2x\cos 3x = \frac{1}{2}\int 2\sin 2x\cos 3x$

$$\frac{1}{2}\int[\sin x + \sin x(-x)]dx$$

$\frac{1}{2}\int (\sin 5x - \sin x)\, dx$

$\frac{1}{2}\left[\frac{-\cos 5x}{5} - \left(\frac{-\cos x}{1}\right)\right] + c$

$\frac{1}{2}\left[\frac{-\cos 5x}{5} + \cos x\right] + c$

$= \frac{-\cos 5x}{10} + \frac{\cos x}{2} + c$

$= \frac{1}{2}\cos x - \frac{1}{10}\cos 5x + c$

Now from (1)

$x\left[\frac{1}{2}\cos x - \frac{1}{10}\cos 5x\right] - \left[\int \left(\frac{1}{2}\right)\cos x - \frac{1}{10}\cos(5x)\right] dx$

$x\left(\frac{1}{2}\cos x - \frac{1}{10}\cos 5x\right) - \left[\int \frac{1}{2}\cos x dx - \int \frac{1}{10}\cos 5x dx\right]$

$x\left(\frac{1}{2}\cos x - \frac{1}{10}\cos 5x\right) - \left[\frac{1}{2}\sin x - \frac{1}{10}\frac{\sin 5x}{5}\right] + c$

$\frac{1}{2}x\cos x - \frac{1}{10}x\cos 5x - \frac{1}{2}\sin x + \frac{\sin 5x}{50} + c$

14. $\int x^2 \sin x dx$

Solution

$$\int uvdx = u\int vdx - \int \left[\left(\frac{du}{dx}\right)\int vdx\right] dx$$

$x^2 \int \sin x dx - \int \left[\frac{d}{dx}(x^2)\int \sin x dx\right] dx$

$x^2(-\cos x) - \int [2x(-\cos x)]\, dx$

$-x^2 \cos x - \int [-2x\cos x]\, dx$

$$-x^2 \cos x + \int 2x \cos x dx$$

$$-x^2 \cos x + 2\int x \cos x dx$$

$$-x^2 \cos x + 2\left[x.\int \cos x dx - \int \{\frac{d}{dx}(x)\int \cos x dx\}dx\right]$$

$$-x^2 \cos x + 2\left[x \sin x - \int \sin x dx\right]$$

$$-x^2 \cos x + 2x \sin x + 2\cos x + C$$

$$2x \sin x - x^2 \cos x + 2\cos x + C$$

15. $\int x^2 \cos 2d \;\; dx$

Solution

$$\int \boldsymbol{uvdx} = \boldsymbol{u}\int \boldsymbol{vdx} - \int \left[\left(\frac{\boldsymbol{du}}{\boldsymbol{dx}}\right)\int \boldsymbol{vdx}\right]\boldsymbol{dx}$$

$$\boldsymbol{\cos^2 x = \frac{1 + \cos 2x}{2}}$$

$$\int x^2 \left(\frac{1 + \cos 2x}{2}\right)dx = \int (\frac{x^2}{2} + \frac{x^2}{2}\cos 2x)dx$$

$$\int \frac{x^2}{2}dx + \int \frac{x^2}{2}\cos 2x dx$$

$$\frac{1}{2}\int x^2 dx + \frac{1}{2}\int x^2 \cos 2x dx$$

$$\frac{1 * x^3}{2 \;\; 3} + \frac{1}{2}\left[x^2 \int \cos 2x dx - \int \{\frac{d}{dx}(x^2)\int \cos 2x dx\}dx\right]$$

$$\frac{x^3}{6} + \frac{1}{2}\left[x^2 \frac{\sin 2x}{2} - \int \left[2x \frac{\sin 2x}{2}\right]dx\right]$$

$$\frac{x^3}{6}+\frac{1}{2}\left[x^2\frac{\sin 2x}{2}-\int x\sin 2x dx\right]$$

$$\frac{x^3}{6}+\frac{1}{2}\left[x^2\frac{\sin 2x}{2}-\left(x\frac{-\cos 2x}{2}-\int-\frac{\cos 2x}{2}\right)dx\right]$$

$$\frac{x^3}{6}+\frac{1}{2}\left[x^2\frac{\sin 2x}{2}+x\frac{\cos 2x}{2}-\frac{1}{2}\frac{\sin 2x}{2}\right]+C$$

$$\frac{x^3}{6}+\frac{x^2\sin 2x}{4}+\frac{x\cos 2x}{4}-\frac{\sin 2x}{8}+C$$

16. $\int x\tan^{-1}x$

Solution

$$=\tan^{-1}[x\int xdx-\int\left\{\frac{d}{dx}](\tan^{-1}x)\int xdx\right\}dx$$

$$=\tan^{-1}x*\frac{x^2}{2}-\int\left[\frac{1}{(1+x^2)}*\frac{x^2}{2}\right]dx$$

$$=\tan^{-1}x*\frac{x^2}{2}-\frac{1}{2}\int\frac{1}{(1+x^2)}*x^2\,dx$$

$$=\tan^{-1}x*\left(\frac{x^2}{2}\right)-\frac{1}{2}\int\frac{x^2}{1+x^2}dx$$

$$=\frac{x^2}{2}\tan^{-1}x-\frac{1}{2}\int\frac{x^2+1-1}{1+x^2}dx$$

$$=\frac{x^2}{2}\tan^{-1}x-\frac{1}{2}\int[\left(\frac{1+x^2}{1+x^2]}-\frac{1}{1+x^2}\right)dx$$

$$=\frac{x^2}{2}\tan^{-1}x-\frac{1}{2}\int\left(1-\frac{1}{1+x^2}\right)dx$$

$$= \frac{x^2}{2}\tan^{-1}x - \frac{1}{2}\left[\int 1dx - \int \frac{1}{1+x^2}dx\right]$$

$$= \frac{x^2}{2}\tan^{-1}x - \frac{1}{2}[x - \tan^{-1}x] + C$$

$$= \frac{x^2}{2}\tan^{-1}x - \frac{1}{2}x + \tan^{-1}\frac{x}{2} + C$$

$$\frac{1}{2}[x^2 + \tan^{-1}x + \tan^{-1}x - x +] + C$$

17. $\int x\frac{\sin^{-1}x}{\sqrt{1-x^2}}dx$

Solution

Let $z = \sin^{-1}x$ (so, x=sinz)

$$\frac{dz}{dx} = \frac{1}{\sqrt{1-x^2}}$$

$$dz = \frac{dx}{\sqrt{1-x^2}}$$

$$\int x\frac{\sin^{-1}x}{\sqrt{1-x^2}}dx = \int z.\sin z * \frac{dx}{\sqrt{1-x^2}}$$

$$= \int z.\sin z.dz$$

$$z\int \sin zdz - \left[\int \{\frac{d}{dz}(z)\int \sin zdz\}dz\right]$$

$$z(-\cos z) - \int -\cos z\, dz$$

$$-z\cos z + \int \cos zdz$$

$$-z\cos z + \sin z + C \text{---------(1)}$$

As, $\sin z = x$

$\sin^2 z = x^2$

$$(1-\cos^2 z) = x^2$$

$$\cos^2 z = 1 - x^2$$

$$\mathbf{\cos z = \sqrt{1-x^2}}$$

So required Integration from (1)

I= $-(\sin^{-1} x) * \sqrt{1-x^2} + x + C$

18. $\int \mathbf{\sec^3 x dx}$

Solution

$$\int \mathbf{\sec^3 x dx} = \int \mathbf{secx.sec^2 x dx}$$

$$\boldsymbol{Use \int uvdx = u \int vdx - \int \left[\left(\frac{du}{dx}\right) \int vdx\right] dx}$$

$$= \sec x \tan x - \int \sec x \tan^2 x dx$$

$$\sec x \tan x - \int \sec x (\sec^2 x - 1) dx$$

$$\sec x \tan x - \int \sec^3 x dx + \int \sec x dx$$

$$\int \mathbf{\sec^3 x dx} \quad = \sec x \tan x - \int \sec^3 dx + \int \sec x dx$$

$$\int (\sec^3 x + \sec^3 x)\, dx = \text{secx.} \tan x + \int \sec x dx$$

$$2 \int \sec^3 x\, dx = \sec x \tan x + \log(\sec x + \tan x) + C$$

$$\int \mathbf{\sec^3 x dx} = \frac{1}{2}[\sec x \tan x + \log(\sec x + \tan x) + C]$$

DEFINITE INTEGRALS

Properties

$$\int_a^b f(x)dx = -\int_b^a f(x)dx$$

$$\int_a^b f(x)dx = \int_a^c f(x)dx + \int_a^b f(x)dx$$ (where a<c<b)

$$\int_a^b f(x)dx = \int_a^b f(a+b-x)dx$$

$$\int_0^a f(x)dx = \int_0^a f(a-x)dx$$

$$\int_a^b f(x)dx = [f(x)]_a^b = f(b) - f(a)$$

1. $\int_0^{\frac{\pi}{2}} \cos x dx$

$\int \cos x dx = \sin x + C$

$[\sin x + C]_0^{\frac{\pi}{2}}$

$\int f(x)dx = [f(x)]_a^b$

$= f(b) - f(a)$

$= \sin\frac{\pi}{2} + C - (\sin 0 + C)$

$\sin\frac{\pi}{2} = 0, \sin 0 = 0$

$= 1$

2. $\int_0^1 (4x^3 + 3x^2 - 2x + 1)dx$

$$\int \left((4x^3 + 3x^2 - 2x + 1)dx\right) = 4\frac{x^4}{4} + 3\frac{x^3}{3} - 2\frac{x^2}{2} + x$$

$$\int_0^1 (4x^3 + 3x^2 - 2x + 1)dx = [x^4 + x^3 - x^2 + x]_0^1$$

$$= [1 + 1 - 1 + 1 - (0 + 0 - 0 + 0)]_2$$

3. $$\int_0^{\frac{\pi}{4}} \frac{\sin x}{\cos 3x + 3\cos x} dx$$

$$\cos 3x = 4\cos^3 x - 3\cos x$$

$$\cos 3x + 3\cos x = 4\cos^3 x$$

$$\int_0^{\frac{\pi}{4}} \frac{\text{Sin}\, x}{4\,\text{Cos}^3 x} dx = \frac{1}{4}\int_0^{\frac{\pi}{4}} \frac{\text{Sin}\, x}{\text{Cos}\, x} * \frac{1}{\text{Cos}^2 x} dx$$

$$\frac{1}{4}\int_0^{\frac{\pi}{4}} \text{Sec}^2 x \tan x\, dx$$

$$= \frac{1}{4}\int_0^{\frac{\pi}{4}} \sec x(\sec x \tan x)\, dx$$

$$\sec x = t$$

$$(\sec x . \tan x)dx = dt$$

upper limit $\sec\frac{\pi}{4} = \sqrt{2}$

lower limit $\sec 0 = 1$

$$\frac{1}{4}\int_1^{\sqrt{2}} t dt = \frac{1}{4}\left[\frac{t^2}{2}\right]_1^{\sqrt{2}}$$

$$= \frac{1}{8}[(\sqrt{2})2 - (1)2]$$

$$= \frac{1}{8}(2 - 1)$$

$$\frac{1}{8}$$

4. $$\int_0^{\frac{\pi}{4}} \tan^2 x dx$$

$$1 + \tan^2 x = \sec^2 x$$

$$\tan^2 x = \sec^2 x - 1$$

$$I = \int_0^{\frac{\pi}{4}} (\sec^2 x - 1)\, dx$$

$$I = [\tan x]_0^{\frac{\pi}{4}} - [x]_0^{\frac{\pi}{4}}$$

$$I = \left[\tan\frac{\pi}{4} - \tan 0\right] - \left(\frac{\pi}{4} - 0\right)$$

$$= (1 - 0) - \left(\frac{\pi}{4}\right)$$

$$= 1 - \frac{\pi}{4}$$

$$= 4 - \frac{\pi}{4}$$

5. $\int_0^{\frac{\pi}{2}} \frac{dx}{1 + \sin x}$

$$I = \int_0^{\frac{\pi}{2}} \frac{dx}{1 + \sin x}$$

$$\int_a^b f(x)\,dx = \int_a^b f(a + b - x)\,dx$$

$$I = \int_0^{\frac{\pi}{2}} \frac{dx}{1 + \sin\left(\frac{\pi}{2} - 0 - x\right)}$$

$$= \int_0^{\frac{\pi}{2}} \frac{dx}{1 + \cos x}$$

$$1 + \cos 2x = 2\cos^2 x$$

$$1 + \cos x = 2\cos^2 \frac{x}{2}$$

$$= \int_0^{\frac{\pi}{2}} \frac{dx}{2\cos^2\frac{x}{2}}$$

$$= \frac{1}{2}\int_0^{\frac{\pi}{2}} \sec^2\frac{x}{2}dx$$

$$= \frac{1}{2}\left[\frac{\tan\frac{x}{2}}{\frac{1}{2}}\right]_0^{\frac{\pi}{2}}$$

$$= \frac{1}{2} * 2\left[\tan\frac{x}{2}\right]_0^{\frac{\pi}{2}}$$

$$\left[\tan\frac{\pi}{4} - \tan 0\right]$$

1-0= 1

6. $\int_0^{\frac{\pi}{2}} \frac{\cos x}{(1 + \sin x)2}$

Let $\text{Sin}\, x = t$

$\text{Cos}\, x dx = dt$

$$\int_0^1 \frac{dt}{1 + t^2}$$

$$[\tan^{-1} t]_0^1$$

$$\tan^{-1} 1 - \tan^{-1} 0$$

$$\tan^{-1}\left(\tan\frac{\pi}{4}\right) - \tan\,(\tan - 0)$$

$$= \frac{\pi}{4}$$

7. $\int_0^{\frac{\pi}{4}} \frac{\tan^3 x}{(1+\cos 2x)} dx$

$= \int_0^{\frac{\pi}{4}} \frac{\tan^3 x}{2\,\mathrm{Cos}^2 x} dx$

$= \frac{1}{2}\int_0^{\frac{\pi}{4}} \tan^3 x \sec^2 x dx$

Let $\tan x = t \tan x = t$

$\sec^2 x dx = dt$

$\frac{1}{2}\int_0^{\frac{\pi}{4}} t^3 dt$

$= \frac{1}{2}\left[\frac{t^4}{4}\right]_0^1$

$= \frac{1}{2}\left[\frac{1}{4} - 0\right]$

$= \frac{1}{8}$

8. $\int_0^{\pi} \boldsymbol{x \sin^2 x dx}$

$I = \int_0^{\pi} x \sin^2 x dx$

$\mathbf{\cos 2x = 1 - 2\sin^2 x}$

$\sin^2 x = \frac{1-\cos 2x}{2}$

$I = \int_0^{\pi} x\left(\frac{1-\cos 2x}{2}\right)dx$

$= \frac{1}{2}\int_0^{\pi} x dx - \frac{1}{2}\int_0^{\pi} x \cos 2x dx$

$\left[\int (uv)dx = u\int vdx - \int \frac{d}{dx}u \int vdx\right] dx$

$$I = \frac{1}{2}\left(\frac{x^2}{2}\right)_0^\pi - \frac{1}{2}\left[x\frac{\sin x}{2} - \int \frac{\sin 2x}{2}dx\right]_0^\pi$$

$$I = \frac{1}{2}\frac{\pi^2}{2} - \frac{1}{4}[x \sin 2x]_0^\pi - \frac{1}{2}\left[\frac{\cos 2x}{2}\right]_0^\pi$$

$$I = \frac{\pi^2}{4} - \frac{1}{4}[0] - \frac{1}{4}[\cos 2\pi - \cos 0]$$

$\cos 2\pi = 1, \cos 0 = 1$

$= 0 - 0$

$$I = \frac{\pi^2}{4} - \frac{1}{4}$$

9. $\int_0^{\frac{\pi}{2}} e^x \cos x dx$

$$I = \int_0^{\frac{\pi}{2}} e^x \cos x dx$$

$$= [\cos x(e^x)]_0^{\frac{\pi}{2}} - \int_0^{\frac{\pi}{2}} (-\sin x)e^x dx$$

$$\left[\cos\frac{\pi}{2} e^{\frac{\pi}{2}} - \cos 0 e^0\right] + \int_0^{\frac{\pi}{2}} e^x \sin xdx$$

$$= [0 - 1] + [e^x \sin x]_0^{\frac{\pi}{2}} - \int \cos xe^x\, dx$$

$$2I = [0 - 1] + \left[e^{\frac{\pi}{2}} \sin\frac{\pi}{2} - e^0 \sin 0\right]$$

$$= -1 + e^{\frac{\pi}{2}}$$

$$I = \frac{e^{\frac{\pi}{2}} - 1}{2}$$

10. $\int_0^e \frac{e^x}{x}[1 + x\log x]dx$

$$= \int e^x [f(x) - f'(x)]dx = e^x f(x)$$

$$= \int e^x \left[\frac{1}{x} + \log x\right] dx$$

$$= [e^x \log x]_1^e = 0$$

$$= e^x \log e$$

11. $\int_0^{\frac{\pi}{2}} \frac{\sqrt{\cos x}}{\sqrt{\cos x} + \sqrt{\sin x}} dx$

$$I = \int_0^{\frac{\pi}{2}} \frac{\sqrt{\cos x}}{\sqrt{\cos x} + \sqrt{\sin x}} dx \quad \text{.........(1)}$$

$$I = \int_0^{\frac{\pi}{2}} \frac{\sqrt{\cos\left(\frac{\pi}{2} - x\right)}}{\sqrt{\cos\left(\frac{\pi}{2} - x\right)} + \sqrt{\sin\left(\frac{\pi}{2} - x\right)}} dx$$

Using properties $\int_0^a f(x)dx = \int_0^a f(a - x)dx$

$$I = \int_0^{\frac{\pi}{2}} \frac{\sqrt{\sin x}}{\sqrt{\sin x} + \sqrt{\cos x}} dx \quad \text{.....(2)}$$

Equations (1+2)

$$2I = \int_0^{\frac{\pi}{2}} \left(\frac{\sqrt{\sin x} + \sqrt{\cos x}}{\sqrt{\sin x} + \sqrt{\cos x}}\right) dx$$

$$2I = \int_0^{\frac{\pi}{2}} dx$$

$$= [x]_0^{\frac{\pi}{2}} = \frac{\pi}{2} - 0$$

$$2I = \frac{\pi}{2}$$

$$I = \frac{\pi}{4}$$

12. $\int_0^{\frac{\pi}{2}} \frac{\sqrt{\tan x}}{1+\sqrt{\tan x}}dx$

$$I = \int_0^{\frac{\pi}{2}} \frac{\sqrt{\tan x}}{1+\sqrt{\tan x}}dx \qquad \text{.............(1)}$$

$$\int_0^{\frac{\pi}{2}} \frac{\sqrt{\tan\left(\frac{\pi}{2}-x\right)}}{1+\sqrt{\tan\left(\frac{\pi}{2}-x\right)}}dx$$

Using properties $\int_0^a f(x)dx = \int_0^a f(a-x)dx$

$$I = \int_0^{\frac{\pi}{2}} \frac{\sqrt{\cot x}}{1+\sqrt{\cot x}}dx$$

$$I = \int_0^{\frac{\pi}{2}} \frac{\sqrt{\cot x}}{1+\sqrt{\left(\frac{1}{tanx}\right)}}dx$$

$$I = \int_0^{\frac{\pi}{2}} \frac{\sqrt{\cot x}.\ \sqrt{\tan x}}{1+\sqrt{\tan x}}dx$$

$$I = \int_0^{\frac{\pi}{2}} \frac{1}{1+\sqrt{\tan x}}dx \qquad \text{.......(2)}$$

Equations (1+2)

$$2I = \int_0^{\frac{\pi}{2}} \frac{1+\sqrt{\tan x}}{1+\sqrt{\tan x}}dx$$

$$2I = \int_0^{\frac{\pi}{2}} dx$$

$$= [x]_0^{\frac{\pi}{2}} = \frac{\pi}{2} - 0$$

$$2I = \frac{\pi}{2}$$

$$I = \frac{\pi}{4}$$

Second Methods

$$\int_0^{\frac{\pi}{2}} \frac{\sqrt{\tan x}}{1+\sqrt{\tan x}}dx = ?$$

$$I = \int_0^{\frac{\pi}{2}} \frac{\sqrt{\frac{\sin x}{\cos x}}}{1+\sqrt{\frac{\sin x}{\cos x}}}dx = \int_0^{\frac{\pi}{2}} \frac{\sqrt{\sin x}}{\sqrt{\cos x}+\sqrt{\sin x}}dx \qquad \text{.......(1)}$$

$$I = \int_0^{\frac{\pi}{2}} \frac{\sqrt{\sin\left(\frac{\pi}{2}-x\right)}}{\sqrt{\cos\left(\frac{\pi}{2}-x\right)}+\sqrt{\sin\left(\frac{\pi}{2}-x\right)}}dx$$

Using properties $\int_0^a f(x)dx = \int_0^a f(a-x)dx$

$$I = \int_0^{\frac{\pi}{2}} \frac{\sqrt{\cos x}}{\sqrt{\sin x}+\sqrt{\cos x}}dx \qquad \text{........(2)}$$

Equations (1+2)

$$2I = \int_0^{\frac{\pi}{2}} \left(\frac{\sqrt{\sin x}+\sqrt{\cos x}}{\sqrt{\sin x}+\sqrt{\cos x}}\right)dx$$

$$2I = \int_0^{\frac{\pi}{2}} dx$$

$$= [x]_0^{\frac{\pi}{2}} = \frac{\pi}{2} - 0$$

$$2I = \frac{\pi}{2}$$

$$I = \frac{\pi}{4}$$

13. Find the area of the region bounded by the line $6x + 5y = 30$,x – axis and the lines $x = -1$ and $x = 3$.

Solution

The region is sketched in the diagram. it lies above the x axis. hence, the required is given by

A= $\int_{-1}^{3} y\,dx = \int_{-1}^{3}\left(\frac{30-6x}{5}\right)dx = [\frac{30x-3x^2}{5}]_{-1}^{3}$

= $\left(\frac{90-27}{5}\right) - \left(\frac{-30-3}{5}\right)$ =

$A = \frac{96}{5}$.

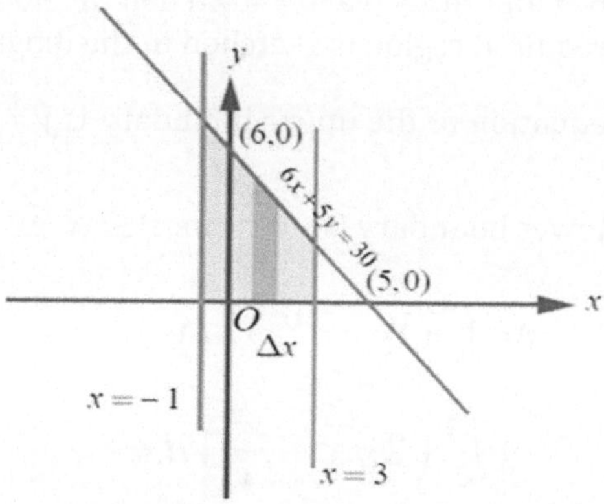

14. Find the area of the region bounded by the line $7x - 5y = 35$, x – axis and the lines $x = -2$ and $x = 3$.

Solution

The region is sketched in the diagram. It lies below the x – axis. Hence, the required area is given by

$A = \left|\int_{-2}^{3} y\,dx\right| = \left|\int_{-2}^{3}\left(\frac{7x-35}{5}\right)dx\right|$

$= \frac{1}{5}\left|\left(7\left(\frac{x^2}{2}\right) - 35x\right)_{-2}^{3}\right|$

$= \frac{1}{5}\left|\left(\left(\frac{63}{2}\right) - 105\right) - (84)\right|$

$A \; = \frac{63}{2}$

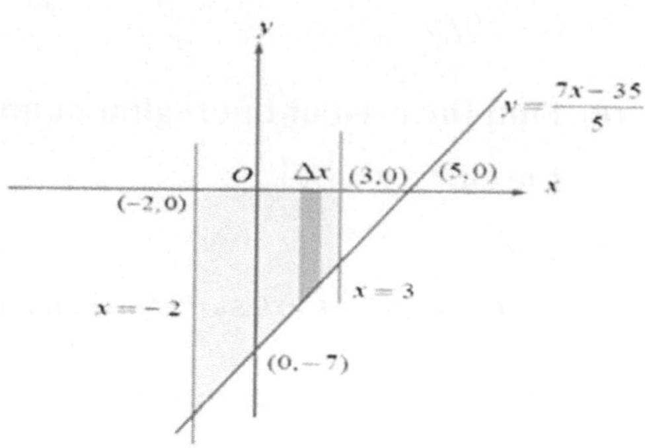

15. Find the area of the region bounded between the parabolas $y^2 = 4x$ and $x^2 = 4y$.

Solution

First, we get the points of intersection of the parabolas. For this, we solve $y^2 = 4x$ and $x^2 = 4y$ simultaneously: Eliminating y between them, we get $x^4 = 64x$ and so $x = 0$ and $x = 4$. Then the points of intersection are (0,0) and (4, 4). The required region is sketched in the diagram Viewing in the direction of y -axis, the equation of the upper boundary is $y = 2\sqrt{x}$ for $0 \leq x \leq 4$ and the equation of the lower boundary is $y = \frac{x^2}{4}$ for $0 \leq x \leq 4$. So the required area Δ is

$$A = \int_0^4 (Y_U - Y_L)\, dx$$

$$= \int_0^4 \left(2\sqrt{x} - \frac{x^2}{4}\right) dx$$

$$= \left[2\left(\frac{2x^{3/2}}{3}\right) - \frac{x^3}{12}\right]_0^4$$

$$= \left[2\left(\frac{2\times8}{3}\right) - \frac{64}{12}\right] - 0$$

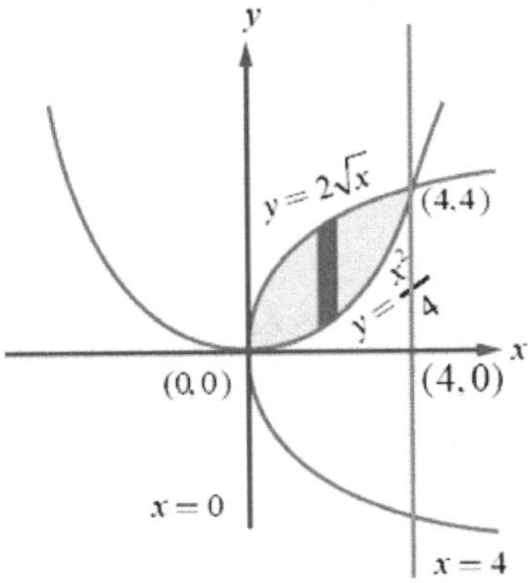

A = 16/3

16. Find the area of the region bounded between the parabola $x^2 = y$ and the curve $y = |x|$.

Solution

Both the curves are symmetrical about y -axis.

The curve $y = |x|$ is $y=\{x \ \ if\, x \geq 0$

$-x \ if\, x \leq 0.$

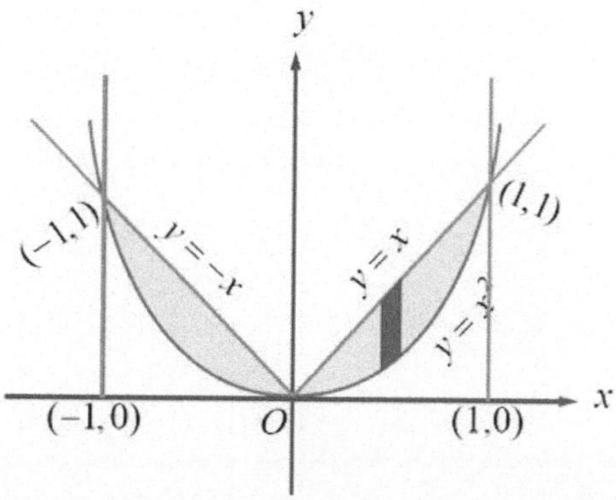

It intersects the parabola $x2 = y$ at (1,1) and (−1,1).

The area of the region bounded by the curves is sketched in diagram. It lies in the first quadrant as well as in the second quadrant. By symmetry, the required area is twice the area in the first quadrant.

In the first quadrant, the upper curve is $y = x, 0 \leq x \leq 1$ and the lower curve is $y = x^2, 0 \leq x \leq 1$.Hence, the required area is given by

$$A = 2\int_0^1 [Y_U - Y_L]\, dx$$

$$= 2\int_0^1 [x - x^2] dx$$

$$= 2\left[\frac{x^2}{2} - \frac{x^3}{3}\right]_0^1$$

$$=2\left(\frac{1}{2} - \frac{1}{3}\right)\left(\frac{1}{2} - \frac{1}{3}\right)$$

$$A = \frac{1}{3}$$

Chapter 7

Partial Differentiation

1. If u = $cos^{-1}\frac{x+y}{\sqrt{x}+\sqrt{y}}$, prove x.$\frac{\partial u}{\partial x}$+ y.$\frac{\partial u}{\partial y}$= -$\frac{1}{2}$cot u

$\Rightarrow$ u = $cos^{-1}\frac{x+y}{\sqrt{x}+\sqrt{y}}$, cos u = $\frac{x+y}{\sqrt{x}+\sqrt{y}}$

Let z = cos u

z = $\frac{x+y}{\sqrt{x}+\sqrt{y}}$ = $\frac{X(1+y/x)}{\sqrt{x}(1+\frac{\sqrt{y}}{\sqrt{x}}}$ = $\frac{x(1+y/x)}{\sqrt{x}(1+(y/x)^{1/2})}$ = $x^{1/2}.\frac{1+y/x}{1+(y/x)^{1/2}}$

z = $x^{1/2}$f(y/x) [homogeneous function of degree 1/2]

By Euler theorem , x.$\frac{\partial z}{\partial x}$+ y.$\frac{\partial z}{\partial y}$ = n × Z ["n" is the degree=1/2]

= n× cos u = $\frac{1}{2}$. cos u

$$x.\frac{\partial(\cos u)}{\partial x} + y.\frac{\partial(\cos u)}{\partial y} = \frac{1}{2}.\cos u$$

$$x.\sin u.\frac{\partial u}{\partial x} + y(-\sin u).\frac{\partial u}{\partial y} = \frac{1}{2}.\cos u$$

$$x.\frac{\partial u}{\partial x} + y.\frac{\partial u}{\partial y} = \frac{-1}{2}\frac{\cos u}{\sin u}$$

$$\mathbf{x.\frac{\partial u}{\partial x} + y.\frac{\partial u}{\partial y} = \frac{-1}{2}\cot u}$$

2. If $u = \sin^{-1}\frac{x^2+y^2}{x-y}$**, prove** $x.\frac{\partial u}{\partial x} + y.\frac{\partial u}{\partial y} = ?$

$$\Rightarrow u = \sin^{-1}\frac{x^2+y^2}{x-y}, \quad \sin u = \frac{x^2+y^2}{x-y}$$

$$\sin u = \frac{x^2(1+\frac{y^2}{x^2})}{x(1-y/x)} = \frac{x(1+\frac{y^2}{x^2})}{1-y/x}$$

$$\sin u = z$$

$$z = x.\frac{x(1+\frac{y^2}{x^2})}{1-y/x}$$

By Euler theorem, $x.\frac{\partial z}{\partial x} + y.\frac{\partial z}{\partial y} = n.z$ (here n = 1)

$$x.\frac{\partial(\sin u)}{\partial x} + y.\frac{\partial(\sin u)}{\partial y} = n.\sin u$$

$$x.\cos u.\frac{\partial u}{\partial x} + y.\cos u.\frac{\partial u}{\partial y} = n.\sin u$$

$$x.\frac{\partial u}{\partial x} + y.\frac{\partial u}{\partial y} = \frac{\sin u}{\cos u} = \tan u$$

$$\mathbf{x.\frac{\partial u}{\partial x} + y.\frac{\partial u}{\partial y} = \tan u}$$

3. $u = tan^{-1}\frac{x^3+y^3}{x+y}$

$$\Rightarrow \tan u = \frac{x^3+y^3}{x+y} = \frac{x^3\,(1+\frac{y^3}{x^3})}{x(1+y/x)}$$

$$= \frac{x^2\,(1+\frac{y^3}{x^3})}{1+y/x}$$ Homogeneous equation of degree 2

z = tan u

By Euler theorem, $x.\frac{\partial z}{\partial x} + y.\frac{\partial z}{\partial y} = n.z$ (here n = 2)

$$x.\frac{\partial(\tan u)}{\partial x} + y.\frac{\partial(\tan u)}{\partial y} = 2.\tan u$$

$$x.\, sec^2\, u.\frac{\partial u}{\partial x} + y.\, sec^2\, u.\frac{\partial u}{\partial y} = 2.\tan u$$

$$x.\frac{\partial u}{\partial x} + y.\frac{\partial u}{\partial y} = \frac{2\tan u}{\sec^2 u}$$

$$= \frac{2\sin u}{\cos u}\cos^2 u$$

$$x.\frac{\partial u}{\partial x} + y.\frac{\partial u}{\partial y} = 2\sin u.\cos u = \sin 2u$$

$$\mathbf{x.\frac{\partial u}{\partial x} + y.\frac{\partial u}{\partial y} = \sin 2u}$$

4. $u = tan^{-1}\frac{x^3+y^3}{x-y}$

$$\Rightarrow \tan u = \frac{x^3+y^3}{x-y} = \frac{x^3\,(1+\frac{y^3}{x^3})}{x(1-y/x)}$$

$$= \frac{x^2\left(1+\frac{y^3}{x^3}\right)}{1-y/x}$$

z = tan u

By Euler theorem, $x.\frac{\partial z}{\partial x} + y.\frac{\partial z}{\partial y} = n.z$

$$x.\frac{\partial(\tan u)}{\partial x} + y.\frac{\partial(\tan u)}{\partial y} = 2.\tan u$$

$$x.\sec^2 u.\frac{\partial u}{\partial x} + y.\sec^2 u.\frac{\partial u}{\partial y} = 2.\tan u$$

$$x.\frac{\partial u}{\partial x} + y.\frac{\partial u}{\partial y} = \frac{2\tan u}{\sec^2 u}$$

$$= \frac{2\sin u}{\cos u}.\cos^2 u$$

$$x.\frac{\partial u}{\partial x} + y.\frac{\partial u}{\partial y} = 2\sin u.\cos u = \sin 2u$$

$$\mathbf{x.\frac{\partial u}{\partial x} + y.\frac{\partial u}{\partial y} = \sin 2u}$$

5. **u =** $x^2\tan^{-1}$ **(y/x) –** $y^2\tan^{-1}$ **(x/y), prove that** $\frac{\partial^2 y}{\partial x.\partial y} = \frac{x^2 - y^2}{x^2 + y^2}$

$$\Rightarrow \frac{\partial u}{\partial y} = \frac{x^2\partial\tan^{-1}\left(\frac{y}{x}\right)}{\partial\left(\frac{y}{x}\right)} \times \frac{\partial\left(\frac{y}{x}\right)}{\partial y} - [\frac{y^2\partial\tan^{-1}\left(\frac{x}{y}\right)}{\partial\left(\frac{x}{y}\right)} \times \frac{\partial\left(\frac{x}{y}\right)}{\partial y}$$

$$+ \tan^{-1}(x/y).2y\,]$$

$$= x^2.\frac{1}{1+\left(\frac{y}{x}\right)^2} * \frac{1}{x} \quad - (y^2.\frac{1}{1+\left(\frac{x}{y}\right)^2}.x\,.\frac{\partial y^{-1}}{\partial y} + \tan^{-1}(x/y).2y\,)$$

$$= \frac{1}{x^2 . x^2 + y^2} \times x^2 \times \frac{1}{x} - (y^2 . \frac{y^2}{x^2 + y^2} \times \frac{x(-1)}{y^2} + tan^{-1}(x/y).2y)$$

$$= \frac{x^3}{x^2 + y^2} + \frac{xy^2}{x^2 + y^2} - 2y \, tan^{-1}(x/y)$$

$$= \frac{x^3 + xy^2}{x^2 + y^2} - 2y \, tan^{-1}(x/y)$$

$$= x - 2y \, tan^{-1}(x/y)$$

Now, $\frac{\partial}{\partial x}\left(\frac{\partial u}{\partial y}\right) = \frac{\partial^2 y}{\partial x.\partial y} = 1 - 2y. \frac{\partial tan^{-1}\left(\frac{x}{y}\right)}{\partial\left(\frac{x}{y}\right)} \times \frac{\partial\left(\frac{x}{y}\right)}{\partial x}$

$$1-2y.[\frac{1}{1+(x/y)^2} * \frac{1}{y}]$$

$$1-[2y.\frac{y^2}{(y^2+x^2)} * \frac{1}{y}]$$

$$= \frac{x^2-y^2}{(x^2+y^2)}$$

= proved

6. Find $\frac{\partial f}{\partial x}$ and $\frac{\partial f}{\partial y}$ for the following function,

(A) $f(x, y) = (x^2 - 1)(y + 2)$

Solution

Differentiate with respect to x in order to find $\frac{\partial f}{\partial x}$ [$y_{constant}$]

$$\frac{\partial f}{\partial x} = \frac{\partial}{\partial x}(x^2 - 1)(y + 2)$$

$$= (y + 2)\frac{\partial}{\partial x}(x^2 - 1)$$

$$= (y + 2)(2x)$$

$$= 2x(y + 2)$$

Similarly,

differentiate with respect to y in order to find $\frac{\partial f}{\partial Y}$ [x $_{\text{constant}}$]

$$\frac{\partial f}{\partial Y} = \frac{\partial}{\partial x}(x^2 - 1)(y + 2)$$

$$= (x^2 - 1)\frac{\partial}{\partial y}(y + 2)$$

$$= (x^2 - 1) \cdot 1$$

$$= (x^2 - 1)$$

(B) $f(x,y,z)=4x^3y^2-e^zy^4+z^3x^2+4y-x^{16}$

Find the value of $\frac{\partial f}{\partial x}, \frac{\partial f}{\partial Y}$ *and* $\frac{\partial f}{\partial Z}$

Solution

Differentiate with respect to x in order to find $\frac{\partial f}{\partial x}$[y and zwill be constant]

$$\frac{\partial f}{\partial x}=12x^2y^2-2z^3x-16x^{15}$$

Differentiate with respect to y in order to find $\frac{\partial f}{\partial Y}$ [x and zwill be constant]

$$\frac{\partial f}{\partial Y}=8x^3y-4e^zy^3+4$$

Differentiate with respect to z in order to find $\frac{\partial f}{\partial z}$ [x and ywill be constant]

$$\frac{\partial f}{\partial z}=-e^zy^4+3z^2x^2$$

(C) $w=\cos(x^2+2y) -e^4x-z^4y+y^3$

find the value of $\frac{\partial f}{\partial x}, \frac{\partial f}{\partial Y}$ *and* $\frac{\partial f}{\partial Z}$?

Answer,

Differentiate with respect to x in order to find $\frac{\partial f}{\partial x}$[y, z$_{constant}$]

$\frac{\partial f}{\partial x}$= −2xsin($x^2$+2y) −$e^4$

Differentiate with respect to y in order to find$\frac{\partial f}{\partial Y}$ [x, z $_{constant}$]

$\frac{\partial f}{\partial Y}$= −2sin($x^2$+2y) −$z^4$+3$y^2$

Differentiate with respect to z in order to find$\frac{\partial f}{\partial z}$ [x, y $_{constant}$]

$\frac{\partial f}{\partial z}$= −4$z^3$y

(D) f(u,v,p,t)=8u^2t^3p+2u^2t+3p^4−v

Find the value of$\frac{\partial f}{\partial u}$,$\frac{\partial f}{\partial v}$ *and* $\frac{\partial f}{\partial t}$?

Solution

Differentiate with respect to u in order to find$\frac{\partial f}{\partial u}$ [v, p,t $_{constant}$]

$\frac{\partial f}{\partial u}$= 16u$t^3$p+4ut

Differentiate with respect to v in order to find$\frac{\partial f}{\partial v}$ [u,p,t $_{constant}$]

$\frac{\partial f}{\partial v}$= -1

Differentiate with respect to t in order to find$\frac{\partial f}{\partial u}$ [v,p,u $_{constant}$]

$\frac{\partial f}{\partial t}$=24$u^2t^2$p+2$u^2$

(E)g(s,t,v)=t^2ln(s+2t)−ln(3v)

Find the value of $\frac{\partial g}{\partial s}$ and $\frac{\partial g}{\partial v}$

Solution

Differentiate with respect to u in order to find $\frac{\partial f}{\partial u}$ [t,v $_{constant}$]

$\frac{\partial g}{\partial s}$=$t^2 \frac{1}{s+2t}$

Differentiate with respect to v in order to find $\frac{\partial f}{\partial v}$ [s,t $_{constant}$]

$\frac{\partial g}{\partial v}$=$\frac{-1}{3v}$

(F) $f(x,y)=x^3y^2-\left(\frac{4y^6}{x^3}\right)$

Find the value of $f'_x f'_y$ **and** f'_{xy}

Solution

1st order partial derivatives;

Differentiate with respect to x [$y_{constant}$]

$f'_x=3x^2y^2+12x^{-4}y^6$

Differentiate with respect to y [$x_{constant}$]

$f'_y=2x^3y-24x-3y^5$

Now let's compute each of the mixed **second order partial derivatives.**

$f'_{xy}=(f'_x)_y=6x^2.2y+72x^{-4}.y^5=12x^2y+72x^{-4}.y^5$(differentiation of f'_x with respect to y)

(G) $A\left(\frac{x}{y}\right) = \cos\left(\frac{x}{y}\right)-x^7y^4+y^{10}$

Find the value of $A'_x A'_y$ **and** A'_{xy}

Solution

Differentiate with respect to x [$y_{constant}$] $A'_x = -\frac{\sin\left(\frac{x}{y}\right)}{y} - 7x^6y^4$

Differentiate with respect to y [$x_{constant}$]

$$A'_y = \left(\frac{x}{y2}\right)\sin\left(\frac{x}{y}\right) - 4x^7y^3 + 10y^9$$

Now let's compute each of the mixed **second-order partial derivatives.**

$A'_{xy} = (A'_{x})_y = \left(\frac{1}{y2}\right)\sin\left(\frac{x}{y}\right) + \left(\frac{x}{y3}\right)\cos\left(\frac{x}{y}\right) - 28x^6y^3$((differentiation of A'_x with respect to y)

(H) $g(u,v)=u^3v^4-2u\sqrt{v^3}+u^6-\sin(3v)$

Find the value of g'_u and g'_v

Solution

Differentiating the equation with respect to u gives, [v $_{constant}$]

$$g'_u = 3u^2v^4 - 2v^{\frac{3}{2}} + 6u^5$$

Differentiating the equation with respect to v gives, [u $_{constant}$]

$$g'_v = 4u^3v^3 - 3u\sqrt{v} - 3\cos(3v)$$

(I) $z=x^2\sin(6y)$

Find the value of $\frac{\partial z}{\partial x}$ and $\frac{\partial z}{\partial y}$

Solution

Diff. with respect to x [$y_{constant}$]

$$\frac{\partial z}{\partial x} = 2x\sin(6y)$$

Diff. with respect to y[$x_{constant}$]

$$\frac{\partial z}{\partial y} = 6x^2\cos(6y)$$

(J) $f(x, y) = e^{x+y+1}$

Find the value of $\frac{\partial f \partial f}{\partial x \partial x}$ and $\frac{\partial f \partial f}{\partial y \partial y}$?

Solution

We can write,

$e^{x+y+1} = e^{x} e^{y} e.$

Now, differentiate with respect to $x[y_{constant}]$

$$\frac{\partial f}{\partial x} = \frac{\partial}{\partial x}(e\, e^{x}\, e^{y})$$

$$= e\, e^{y}(e^{x})$$

$$= e^{x+y+1}$$

Similarly,

Now, differentiate with respect to $y\ [x_{constant}]$

$$\frac{\partial f}{\partial y} = \frac{\partial}{\partial y}(e\, e^{x}\, e^{y})$$

$$= e^{x+y+1}$$

Problems based on Euler's method

If $u = \tan^{-1}\left[\frac{X^3+Y^3}{X+Y}\right]$ then find that $x\frac{\partial u}{\partial x} + y\frac{\partial u}{\partial y} = ?$ t

Solution

$$u = \tan^{-1}\left[\frac{X^3+Y^3}{X+Y}\right]$$

$$\tan u = \left[\frac{X^3+Y^3}{X+Y}\right]$$

$$= \frac{X^{3}\left(1+\frac{Y^3}{X^3}\right)}{X\left(1+\frac{Y}{X}\right)}$$

$$= X^2\left[\frac{1+Y^3+X^3}{1+\frac{Y}{X}}\right]$$

z= tan u

by Eulers theorem

$$x\frac{\partial z}{\partial x} + y\frac{\partial z}{\partial y} = nz$$ here "n" is degree = 2

$$x\frac{\partial(tanu)}{\partial x} + y\frac{\partial(\tan u\ )}{\partial y} = 2tanu$$

$$x(\sec 2\,u)\frac{\partial u}{\partial x} + y(sec2u)\frac{\partial u}{\partial y} = 2tanu\frac{\partial u}{\partial x} + y\frac{\partial u}{\partial y}$$

$$= 2\frac{tanu}{sec2u}\frac{\partial u}{\partial x} + y\frac{\partial u}{\partial y} = 2\frac{sinu}{cosu}\ cos2u$$ = 2 sinu cos u

$$\therefore x\frac{\partial u}{\partial x} + y\frac{\partial u}{\partial y}$$ = sin2u

DIFFERENTIAL EQUATION-BASED ON LINEAR EQUATION

$\frac{dy}{dx} + PY = \ Q$ Where P and Q are function in x

Integrating Factor = $e^{\int p.dx}$

Solution y × I.F. = ∫ Q. × I.F.

1. $\frac{dy}{dx}$ **+ y cot x = 2 cos x**

⇒ This is of the form, $\frac{dy}{dx}$+ PY $=\ Q$

Integrating Factor = $e^{\int p.dx} = e^{\int cot\ x.dx} = e^{loge\ sin\ x}$ = sin x

y× I.F. = ∫ Q. × I.F.

y× sin x = ∫ 2cos x. sin x

y× sin x = ∫ sin 2x.dx

$$y.\sin x = -\frac{\cos 2x}{2}$$

$$y = -\frac{\cos 2x}{2\sin x} + c$$

2. $(x+2y^3)\frac{dy}{dx} = y$

$$\Rightarrow \frac{dy}{dx} = \frac{y}{(x+2y^3)} \Rightarrow \frac{dy}{dx} - \frac{y}{(x+2y^3)} = 0$$

Or $\frac{dx}{dy} = \frac{x}{y} + 2y^2$

$\frac{dy}{dx} + P_y = Q$; Integrating Factor = $e^{\int p.dx}$;

I.F. = $e^{\int -1/y.dy} = e^{-\log y} = e^{\log y^{-1}} = \frac{1}{y}$

Solution x × I.F. = ∫ Q. × I.F.

$$x \times \frac{1}{y} = \int 2y^2 \times \frac{1}{y}$$

$$\frac{x}{y} = \frac{2y^2}{2}$$

$$x = y^3 + c$$

3. **Find the general solution of the differential equation xdy -(y + 2x²).dx=0**

Solution

$xdy - (y + 2x^2).dx = 0$

$\frac{dy}{dx} - y/x = 2x$

Comparing this with the differential equation $\frac{dy}{dx} + Py = Q$,we have the values of P = -1/x and the value of Q = 2x.

Integrating factor(IF) = $e^{\int P.dx}$

$e^{\int(-1/x).dx}$

= $e^{-\log x}$

=-1/x

Solution will be

$$y\times IF = \int Q \times IF\, dx$$

$$y\left(\frac{1}{x}\right) = -\int 2x.\frac{1}{x}.dx + c$$

$$\frac{y}{x} = -2\int dx + c$$

$$= -2x + c$$

$$\therefore y = 2x^2 + xc$$

4. Find the general solution of the differential equation

dy/dx + (sec x)y = 7

Solution

Comparing the given equation with dy/dx + Py = Q

We see, P = sec x, Q = 7

Now let's find out the integrating factor using the form

esecxdx

=secx+tanx

Solution of Differential Equation

$$y\times IF = \int Q.IF dx + c$$

$$y(\sec x+\tan x) = \int 7.(\sec x + \tan x)dx$$

$$y(\sec x+\tan x) = 7((\sec x + \tan x) + \log \sec x)$$

$$\therefore y = \frac{7\ln((\sec x+\tan x)+\log \sec x)}{\sec x+\tan x}$$

5. **Find the general solution of the differential equation**

dy/dx + 2y = sin x

Solution

Comparing this with the differential equation $\frac{dy}{dx}$ + Py = Q ,we have the value

$$\frac{dy}{dx} + 2y = sinx$$

We get,

P=2 ,Q=Sinx

IF=$e^{\int pdx}$ =$e^{\int 2dx}$

IF=e^{2x}

Solution of differential Equation

y× e^{2x}=$\int sinx.e^{2x}dx + c$- ----(i)

I=$\int sinx.e^{2x}dx$

I=sinx$\int e^{2x}dx - \int\left[\frac{d(sinx)}{dx}\int e^{2x}dx\right]$dx

I=(sinx) $\frac{e^{2x}}{2} - \int(cosx)\frac{e^{2x}}{2}dx$

I= (sinx) $\frac{e^{2x}}{2} - \frac{1}{2}\left[cosx\int\frac{e^{2x}}{2} - \int -sinx.\int\frac{e^{2x}}{2}\right]$

I= (sinx)$\frac{e^{2x}}{2} - \frac{1}{2}\left[\frac{(cosx)e^{2x}}{4} + \int(sinx.)e^{2x.}\frac{1}{4}\right]$+c

I= (sinx) $\frac{e^{2x}}{2} - \frac{1}{8}(cosx).e^{2x} - \frac{1}{8}I$+C

$$\frac{9I}{4} = \frac{e^{2x}}{8}[4sinx - cosx] + c$$

II=$\frac{e^{2x}}{18}[4sinx - cosx] + c$

Now, Putting I Value in equation 1,

ye^{2x} $= \frac{e^{2x}}{18}[4sinx - cosx] + c$

y=$\frac{1}{18}[4sinx - cosx] + c$

6. **Find the general solution of the differential equation**

$$\frac{dy}{dx} - y = (x-1)$$

Solution

$$\frac{dy}{dx} - y = (x-1)e^x$$

Comparing this with the differential equation $\frac{dy}{dx}$ + Py = Q ,we have the values

P= - 1 Q= $(x-1)e^x$

IF = $e^{\int pdx} = e^{\int(-1)dx} = e^{-x}$

Solution of differential Equation

$$y\times IF = \int Q \times IF\, dx$$

$$y \times e^{-x} = \int (x-1)e^x \times e^{-x} dx$$

$$y \times e^{-x} = \int (x-1)dx$$

$$ye^{-x} = \frac{x^2}{2} - x + c$$

7. **Find the general solution of the differential equation**

$$\frac{dy}{dx} + y = e^{-x}$$

Solution

Comparing this with the differential equation $\frac{dy}{dx}$ + Py = Q ,we have the values

P=1 ,Q=e^{-x}

IF=$e^{\int pdx} = e^{\int 1.dx}$

=e^x

Solution of differential Equation

$y.e^{x} = \int 1.dx + c$

$ye^{x} = x + c$

Homogenous function,

8. $(y^2+xy)dx - x^2dy = 0$

Solution

Put into standard form:

$$\frac{dy}{dx} = \frac{y^2 + xy}{x^2} \; y^2\left(1 + \frac{x}{y}\right)$$

$$y^2\left(\frac{x^2}{y^2}\right)$$

Put $y = v.x$ and $\frac{dy}{dx} = v + v\frac{dv}{dx}$

$V + x\frac{dv}{dx} = -\frac{(v+2v^2)}{1+v}$

$= \frac{-v^2}{1+v}$

$x\frac{dv}{dx} = \frac{(v+2v^2)}{1+v}$

$$\frac{1}{v} - \frac{1}{1+2v}dv = -\frac{dx}{x}$$

Taking integration on both side,

$$\int \left(\frac{1}{v} - \frac{1}{1+2v}\right)dv = -\int \frac{dx}{x}$$

$\log v - \frac{1}{2}\log(1+2v) = \log x + \log c$

$$log \frac{v}{\sqrt{1+2v}} = \log\frac{c}{x}$$

Replace,

$$v = \frac{y}{x}$$

$$\frac{y\sqrt{x}}{\sqrt{x+2y}} = c$$

$$\therefore \frac{y^2 x}{x+2y} = c^2$$

9. $(X+Y)\,dx + (Y - X)\,dy = 0$

Solution

$$\frac{dy}{dx} = \frac{X+Y}{-(Y-X)}$$

Divided by x ,

$$\frac{dy}{dx} = \frac{1+\frac{Y}{X}}{1-\frac{Y}{X}}$$

$$\frac{dy}{dx} = f\left(\frac{Y}{X}\right)$$

We can take $\frac{Y}{X} = v$

$$\frac{dy}{dx} = V\frac{dx}{dx} + x\frac{dv}{dx}$$

$$\frac{v - v^2 - 1 - v}{1 - v} = -x\frac{dv}{dx}$$

$$\frac{v^2+1}{1-v} = x\frac{dv}{dx}$$

Take integral on both side,

$$\int \left[\frac{1-v}{v^2+1}\right] dv = \int \frac{dx}{x}$$

$$\int \frac{1}{v^2+1} dv - \int \frac{v}{v^2+1} dv = logx$$

$$\tan^{-1} v - \int \frac{v}{v^2+1} dv = logx \quad \text{....(1)}$$

Take z= v^2+1

$$\frac{dz}{dv} = 2v$$

Vdv= $\frac{dz}{2}$

$$\text{so required solution(from 1)} = \tan^{-1} v - \int \frac{1}{2}\frac{dz}{z} = logx$$

$$\tan^{-1} v - \frac{1}{2}\log(v^2+1) = logx + c$$

$$\tan^{-1}\frac{y}{x} - \frac{1}{2}\log\left(\frac{y^2}{x^2}+1\right) = \text{logx+c}$$

10. $\mathbf{xy^2\, dy - (x^3 + y^3)\, dx = 0}$

Solution

$$xy^2\, dy = (x^3+y^3)\, dx$$

$$\frac{dy}{dx} = \frac{x^3+y^3}{xy^2} = \frac{x^3[1+(y/x)^3]}{x^2\left[\frac{y^2}{x^2}\right]} \quad \text{(Homogeneous Function)}$$

$$let\ y = vx$$

$$\frac{dy}{dx} = v + x\frac{dv}{dx}$$

Replacing value of $\frac{dy}{dx}$

$$v + x\frac{dv}{dx} = \frac{x^3 + y^3}{xy^2}$$

$$v + x\frac{dv}{dx} = \frac{x^2}{y^2} + v$$

$$x\frac{dv}{dx} = \frac{x^2}{y^2}$$

$$x\frac{dv}{dx} = \frac{1}{\frac{y^2}{x^2}} = \frac{1}{v^2}$$

$$v^2 dv = \frac{dx}{x}$$

take integral on both side

$$\int v^2 dv = \int \frac{dx}{x}$$

$$\frac{v^3}{3} = \log x + c$$

$$v^3 = 3\log x + c$$

$$\frac{y^3}{x^3} = 3\log x + c$$

Chapter 8

Differential Equations Related to Economics

1. **Monthly margin cost for the product is MC = X+30. If the fixed cost is $50, find a cost function?**

$\Rightarrow$ X = Quantity,

$$MC = \frac{dc}{dQ} = \frac{dc}{dx}$$

Here c and Q(=x) represent cost per quantity and total quantity respectively

$$\frac{dc}{dx} = X + 30$$

dc = (30+X). dx

$$\int dc = \int (X + 30.)\, dx$$

$C = X^2 + 30X + K$, K = Integration Constant

Fixed Cost = 50, X = 0

$$C = \frac{0}{2} + 30(0) + K$$

.... K = 50

$$C = \frac{X^2}{2} + 30 + 50$$

2. **Marginal revenue function for product is MR = 44-5X. MC is MC = 3X + 20 & cost of producing 80 units is $11.**

   a) **Find a profit function**

   b) **Find a profit/loss for selling 100 units.**

$$\Rightarrow \quad MR = \frac{dR}{dx}$$

$$\frac{dR}{dx} = 44\text{-}5X$$

$$dR = (44\text{-}5X).dx$$

$$\int dR = \int (44\text{-}5X.)\, Dx$$

$$R = 44X - \frac{5X^2}{2} + K_1$$

$$MC = 3X + 20$$

$$\frac{dc}{dx} = 3X + 20$$

$$dc = 3(X + 20.)\, dx$$

$$\int dc = \int (3X + 20.)\, dx$$

$$c = \frac{3X^2}{2} + 20X + K_2$$

Revenue = 0, When quantity = 0

$$0 = 44(0) - \frac{5(0)^2}{2} + K_1$$

$$K_1 = 0$$

$$R = 44X - \frac{3X^2}{2}$$

$$C = \frac{3X^2}{2} + 20X + K_2$$

Profit = revenue - cost

$$= 44X - \frac{5X^2}{2} - \frac{3X^2}{2} - 20X + 200$$

$= 24X - 4x^2 - 200$

Profit/Loss of selling 100 units, $24(100) - 4(100)^2 - 200$

= $ -37800

Loss = 37,800

3. **Average cost (C) of a product changes at rate of z(x) =** $-6x^{-2} + \frac{1}{6}$. **AC of 6 units is 10, Find AC of 12 units.**

$$\Rightarrow z(x) = -6x^{-2} + \frac{1}{6}$$

$$\frac{dc}{dx} = -6x^{-2} + \frac{1}{6}$$

$$dc = \left(-6x^{-2} + \frac{1}{6}\right).dx$$

$$\int dc = \int \left| -6x^{-2} + \frac{1}{6} \right| .dx$$

$$C = \frac{-6x^{-2+1}}{-2+1} + \frac{1}{6}x + K$$

$$= 6x^{-1} + \frac{1}{6}x + K$$

Given

C = 10, x = 6

$$10 = 6(6^{-1}) + \frac{1}{6} \times 6 + K$$

$10 = 1 + 1 + K$

.......$K = 8$

To find AC of 12 units, (or n = 12)

$$C = 6\times \frac{1}{12} + \frac{1}{6} \times 12 + K$$

$$= \frac{1}{2}(2) + 8$$

C = 10.

CONSUMER SURPLUS OR PRODUCER SURPLUS

4. At a certain factory, MC is $3(q-4)^2$ dollar/unit where level of production is q units. By how much will the total manufacturing cost increase if the production level arises from 6 to 10 units.

$$\Rightarrow \frac{dc}{dq} = 3(q-4)^2$$

$$dc = 3(q-4)^2 .dq$$

$$\int dc = 3(q-4)^2 .dq$$

$$c = 3\int q^2 - 8q + 16.\, dq$$

$$c = \frac{3q^3}{3} - \frac{24q^2}{2} + 48q$$

Production level 6 to 10 units.

$$c = \frac{3q^3}{3} - \frac{24q^2}{2} + 48q$$

$$c = (1000-22\times 100 + 480) - (216 - 432 + 288)$$

$$c = 280-72$$

$$c = 208$$

Total manufacturing cost will increase to dollar 208 if production level rises from 6 to 10 units

5. For a certain item, demand curve is P=D(q) = $\frac{20}{q+1}$ and the supply curve is P=3(q) = q+2. Find the equilibrium price & quantity, then compute consumer & producer surplus.

At equilibrium, D(q) = S(q)

$$\frac{20}{q+1} = q+2$$

$$20 = (q+2)(q+1) = q^2 + 2q + q + 2$$

$$q^2 + 3q - 18 = 0$$

$$(q+6)(q-3) = 0$$

$$q = -6, \; q = 3$$

$$q = 3 \text{ in } D(q) = \frac{20}{q+1} = \frac{20}{3+1} = 20/4 = 5$$

$Pe = 5$

Also qe = 3

Consumer surplus = $\int_0^e D(q).dq - Pe.qe$

$$= \int_0^3 \frac{20}{q+1}.\,dq - 5(3) = 20\int_0^3 \frac{1}{q+1}.\,dq - 15$$

$$= 20\,[log(q+1)]_0^3 - 15$$

$$= 20\,[\log 4 - \log 1] - 15$$

$$= 20\,[2\log 2 - 0] - 15$$

$= \mathbf{12.73}$

Producer surplus = $Pe.qe - \int_0^{qe} S(q).dq$

$$= 15 - \int_0^3 (q+2.)\,dq$$

$$= 15 - [\frac{q^2}{2} + 2q]_0^3$$

$$= 15 - [q/2 + 6/1]$$

$= 15 - [q+12/2]$

$= 15 - 10.5$

$= \mathbf{4.5}$

6. Suppose consumer demand functions for a commodity is D(q) = 4 (25 - q^2) dollars/unit. Find the total amount of money a consumer is willing to spend to get 3 units of commodity.

$\Rightarrow$ $D(q) = 4\,(25 - q^2)$

$$\int_0^3 D(q) = \int_0^3 4\,(25 - q^2).dq$$

$$= 4\left[25q - \frac{q^3}{3}\right]_0^3$$

$$= 4\,[25(3) - \frac{27}{3}]$$

$$= 4\,[\frac{225-27}{3}]$$

$$= \frac{4(198)}{3}$$

$= \mathbf{264}$

Total money consumer willing to spend to get 3 units of commodity is \$264

Chapter 8

Finite Difference and Numerical Analysis

1. If $y = 2x^3 - x^2 + 3x + 1$; calculate the value of Y corresponding to x=0,1,2,3,4,5 and term the table of difference and prove theoretically that second order difference is $12x + 10$, also verify this.

Solution

Difference table:

x	U_x	$\Delta U x$	$\Delta^2 U x$	$\Delta^3 U x$	$\Delta^4 U x$
0	1				
1	5	4			
2	19	14	10		
3	55	36	22	12	
4	125	70	34	12	0
5	241	116	46	12	0

Given

$\Delta^2 U x = 12x + 10$

We know, $\Delta U x = U(x+h) - Ux$

$$= 2(x+h)^3 - (x+h)^2 + 3(x+h) + 1$$
$$- (2x^3 - x^2 + 3x + 1$$

Since h=1

$$\Delta U\, x = 2(x+1)^3 - (x+1)^2 + 3(x+1) + 1 -$$
$$2x^3 + x^2 - 3x - 1$$
$$= 2(x^3 + 3x^2 + 3x + 1) - x^2 - 2x - 1 + 3x + 3 - 2x^3 + x^2 - 3x$$
$$= 2x^3 + 6x^2 + 6x + 2 - x^2 - 2x + 3x + 2 - 2x^3 + x^2 - 3x$$
$$= 6x^2 + 4x + 4$$
$$\Delta^2 U\, x = \Delta(\Delta\, U\, x) = \Delta[U_{(x+h)} - Ux]$$
$$= 6(x+1)^2 + 4(x+1) + 4 - (6x^2 + 4x + 4)$$
$$= 6(x^2 + 1 + 2x) + 4x + 4 + 4 - 6x^2 - 4x - 4$$
$$= 6x^2 + 6 + 12x + 4x + 8 - 6x^2 - 4x - 4$$
$$= 12x + 10$$

If x=0

$$\Delta^2 U_0 = 12 \times 0 + 10 = 10$$

If x=1

$$\Delta^2 U_1 = 12 \times 1 + 10 = 22$$

If x=2

$$\Delta^2 U_2 = 12 \times 2 + 10 = 34$$

If x=3

$$\Delta^2 U_3 = 12 \times 3 + 10 = 46$$

Hence, verified...

PROBLEM BASED ON MISSING TERM

2. Estimate U2 from the following table

x	1	2	3	4	5
Ux	7	?	13	21	37

Since, we are given 4 entries viz; U1, U3, U4, U5, so function can be represented by 3rd degree polynomial.

$\Delta^3 U x = constant$

$\Delta^4 \bigcup x = 0$

Difference table: Let the missing term is 'a'

x	U_x	$\Delta U x$	$\Delta^2 U x$	$\Delta^3 U x$	$\Delta^4 U x$
1	7				
		a-7			
2	a		20-2a		
		13-a		3a-25	
3	13		a-5		38-4a
		8		13-a	
4	21		8		
		16			
5	37				

$$\Delta^4 \bigcup x = 0 \Rightarrow 38 - 4a = 0 \Rightarrow -4a = -38 \Rightarrow a = 9.5$$

(OR)

Δ^4U1=0 E=1+Δ

$(E-1)^4$U1=1

U1$\left[E^4 - 4_{C_1} E^3 + 4_{C_2} E^2 + 4_{C_3} E^1 + 1\right] = 0$

E^4U1$-4_{C_1} E^3$U1+$4_{C_2} E^2$ U1$-4_{C_3} E^1$ U1+ U1=0

U5-4_{C_4}+6U1−4U2+U1=0

$37 - 4 \times 21 + 6 \times 13$ −4U2+7=0

38 −4U2=0

U2=9.5

3. **From the following table of yearly premium for polices maturing at quinquennial ages. Estimate for policy maturing at the ages of 46 years.**

Age(x)	45	50	55	60	65
Premium (Ux)	2.707	2.404	2.083	1.862	1.712

Difference table

Age (x)	Premium ($U x$)	$\Delta U x$	$\Delta^2 U x$	$\Delta^3 U x$	$\Delta^4 U x$
45	2.707	-0.467	0.146	0.046	0.017
50	2.404	-0.321	0.100	-0.029	
55	2.083	0.221	0.070		
60	1.862	0.150			
65	1.712				

Using Newton forward interpolation method:

<u>Formula</u>: $U(a+xh) = Ua + x_{C_1} \Delta Ua + x_{C_2} \Delta^2 Ua + x_{C_3} \Delta^3 Ua + \ldots\ldots.$

(or)

$$U(a+xh) = Ua + \frac{x}{1!}\Delta Ua + \frac{x(x-1)}{2!}\Delta^2 Ua + \frac{x(x-1)(x-2)}{3!}\Delta^3 Ua + \ldots\ldots.$$

46=a+xh

h=5 and a=45

$$U46 = U45 + \frac{1}{5}\Delta U45 + \frac{\left(\frac{1}{5}\right)\left(\frac{1}{5}-1\right)}{2!}\Delta^2 U45 + \frac{\left(\frac{1}{5}\right)\left(\frac{1}{5}-1\right)\left(\frac{1}{5}-2\right)}{3!}\Delta^3 U45$$
$$+\frac{\left(\frac{1}{5}\right)\left(\frac{1}{5}-1\right)\left(\frac{1}{5}-2\right)\left(\frac{1}{5}-3\right)}{4!}\Delta^4 U45$$

$$= 2.707 - 0.0934 - 0.01168 - 0.02208 - 0.5712$$

= 2.0084

NEWTON- GREGORY BACKWARD INTERPOLATION METHODS

4. The following table gives the Census Population of a town for 1961-2001. Estimate the population for the year 1995 by using appropriate interpolation formula.

Years (x)	Population (U_x) (in lakhs)	∇U_x	$\nabla^2 U_x$	$\nabla^3 U_x$	$\nabla^4 U_x$
1961	46	20	-5	2	-3
1971	66	15	-3	-1	
1981	81	12	-4		
1991	93	8			
2001	101				

$$x = \frac{period\ of\ interpotation - last\ argument}{interval\ of\ differencing}$$

$$= \frac{1995-2001}{10}$$

$$= \frac{-6}{10} = -0.6$$

$$U_{b+xh} = U_b + x\Delta\, U_a + \frac{x(x+1)}{2!}\nabla^2\, U\, b + \frac{x(x+1)(x+2)}{3!}\nabla^3\, U\, b + \frac{x(x+1)(x+2)(x+3)}{4!}\nabla^4\, U\, b$$

$$= 101 + (-0.6)8 + \frac{(-0.6)(-0.6+1)}{2}(-4) + \frac{(-0.6)(-0.6+1)(-0.6+2)(-1)}{6} + \frac{(-0.6)(-0.6+1)(-0.6+2)(-0.6+3)}{24}(-3)$$

$$= 101 + 4.8 + 0.48 + 0.056 + 0.1008$$

$$= 101.6368 - 4.8$$

$$= 96.837$$

DIVIDED DIFFERENCES METHODS

5. **The observed value of a function is respectively 168,120,72 and 63 at the 4 position -3,7,9 and 10 of the independent variables. What is the best estimate you can give for the value of the function 6 of the independent variable?**

Divided difference table:

$$U_{a+xh} = U_a + (x-a)\Delta U_a + (x-a)(x-b)\Delta^2 U_a + (x-a)(x-b)(x-c)\Delta^3 U_a$$

$$= 168 + (x-3)(-12) + (x-3)(x-7)(-2) + (x-3)(x-7)(x-9)1$$

$$U_x = 168 - 12x + 36 - [(x^2 - 7x - 3x + 21)(-2)] + [(x-3)(x^2 - 9x - 7x + 63)]$$

$$= 168 - 12x + 36 - 2x^2 + 14x + 6x - 42 + (x^3 - 9x^2 - 7x^2 + 63x - 3x^3 + 27x + 21x - 189)$$

$$= -2x^2 + 8x + 162 + (x^3 - 19x^2 + 111x - 189)$$

$$= x^3 - 2x^2 + 8x + 162 - 19x^2 + 111x - 189$$

$$U_x = x^3 - 21x^2 + 119x - 27$$

$$U_{15} = 15^2(15-1)$$

$$= 225 \times 14$$

$$= 3150$$

By using value of ∂,15 the new divided difference table is

x	U_x	ΔU_x	$\Delta^2 U_x$	$\Delta^3 U_x$
2	4	22	10	1
4	48	52	15	1
5	100	97	21	1
7	294	202	27	1
10	900	310	33	1
11	1210	409	38	
13	2028	561		
15	3150			

6. The numbers of members of certain society are as given in the following table

Years (x)	1990	1991	1992	1993	1994	1995	1996	1997	1998	1999
Members (U_x)	845	867		846	821	772		757	761	796

Make the best estimate you can member in 1992 and 1996

Reference table:

Year x	U_x	ΔU_x	$\Delta^2 U_x$	$\Delta^3 U_x$
1990	845	22	U_3-889	
1991	867	U_3-867	$1713-U_3$	
1992	U_3	$846-U_3$	U_3-871	
1993	846	-25	-24	
1994	821	-49	U_7-723	
1995	772	U_2-772	1529	
1996	U_7	$757-U_7$	U_7-253	
1997	757	4	31	
1998	761	35		
1999	796			

7. From the following table, estimate the number of students who obtained marks between 40 and 45.

Marks (x)	30-40	40-50	50-60	60-70	70-80
No. of students (U_x)	31	42	51	35	31

Difference table:

(x)	(U_x)	ΔU_x	$\Delta^2 U_x$	$\Delta^3 U_x$	$\Delta^4 U_x$
<40	31	42	9	-25	37
<50	73	51	-16	-12	
<60	124	35	-4		
<70	159	31			
<80	190				

$$U_{a+xh}=U_a+\frac{x}{1!}\Delta U_a+\frac{x(x+1)}{2!}\Delta^2 U_a+\frac{x(x+1)(x-2)}{3!}\Delta^3 U_a+\frac{x(x+1)(x-2)(x-3)}{4!}\Delta^4 U_a$$

$$a+xh=45$$

$$40+x\times 10=45$$

$$x=\frac{1}{2}$$

Using Newton Gregory formula, we get

$$U_{45}=U_1\,\frac{x}{1!}\Delta U_1+\frac{x(x-1)}{2!}\Delta^2 U_1+\frac{x(x-1)(x-2)}{3!}\Delta^3 U_1+\frac{x(x-1)(x-2)(x-3)}{4!}\Delta^4 U_1$$

$$=31+\left(\frac{1}{2}\right)\times 42+\frac{(1/2)(1/2-1)}{2}\times 9+\frac{(1/2)(1/2-1)(1/2-2)}{6}\times(-25)+\frac{(1/2)(1/2-1(1/2-3)}{24}\times 37$$

$$=31+21+1.125-1.5625-1.4453$$

$$=47.86$$

48

No. of students gaining marks <45=48

No. of students gaining marks<40=31

No. of students acquiring marks between 40-45 is (48-31) = 17

8. If $h = 1$ then prove that $(E^{-1}\Delta)\,x^3 = 3x^3 - 3x + 1$.

Solution

Given, h=1

LHS $=(E^{-1}\Delta)x^3$

$= \Delta\left(E^{-1}(x^3)\right)$

$= \Delta\,(x-h)^3$

$=\Delta\,(x-h)^3$ $\quad [E^{-1}\,f(x) = f(x-nh)], h = 1]$

$= (x-1+1)^3 - (x-1)^3 \quad [\Delta\, f(x) = f(x+h) - f(x)$

$= x^3 - (x^3 - 3x^2 + 3x - 1)$

$= 3x^2 - 3x + 1$

= R.H.S. = RHs

Hence, it proved

9. If $f(x) = x^2 + 3x$ than show that $\Delta f(x) = 2x + 4$

Solution

Given, f(x)=$x^2 + 3h$

$LHS = \Delta\, f(x)$

$= f(x+h) - f(x)$

$= [(x+h)^2 + 3\,(x+h)] - [x^3 + 3x]$

$= x^2 + h^2 + 2xh + 3x + 3h - x^2 - 3x$

$= h^2 + 2xh + 3h$

$when, h = 1\,, LHS = 1 + 2x + 3$

$= 2x + 4$

$= RHS$

Hence, it proved.

10. Find the missing entry in the following table

x	0	1	2	3	4
y_x	1	3	9	-	81

Solution

Since only four values of f (x) are given, the polynominal which fits the data of degree 3. Hence fourth difference are zero.

$\Delta^4(y_0) = 0$

$$(E-1)^4(y_0) = 0$$

$$(E^4 - 4E^3 + 6E^2 - 4E + 1)(y_0) = 0$$

$y_4 - 4y_3 + 6y_2 - 4y_1 + y_0 = 0$

$$81 - 4\left(y_{3)} + 6(9) - 4(3) + 1\right) = 0$$

$$81 - 4\,y_3 + 54 - 12 + 1 = 0$$

$81 + 54 - 11 = 4y_3$

$$124 = 4y_3$$

$Y_3 = 124/4$

$Y_3 = 31$

11. Following are the population of a district

Year (x)	1881	1891	1901	1911	1921	1931
Population (y) Thousands	363	391	421	-	467	501

Find the population of the year 1911

Solution

Since only five values of f(x) are given, the polynomial which fits the data is od degree 4.

Hence fifth difference are zeroes

$\Delta^5 y_0 = 0$

$$(E-1)^5(y_0) = 0$$

$$(E^5 - 5E^4 + 10E^3 - 10E^2 + 5E - 1)(y_0) = 0$$

$$501 - 5(467) + 10y_3 + 10y_2 + 5y_1 - y_0 = 0$$

$$501 - 5(467) + 10y_3 + 10(421) + 5(391) - 361 = 0$$

$$501 - 2335 + 10y_3 - 10(421) + 1955 - 363 = 0$$

$$10y_3 - 4352 = 0$$

$$10y_3 = 4352$$

$$y_3 = \frac{4352}{4}$$

$$y_3 = 435.2$$

Since the population is given in thousands, population of the year 1911

=435.2 x 1000

=435200

12. Find the missing entries from the following.

x	0	1	2	3	4	5
$y = f(x)$	0	-	8	15	-	35

Solution

Let the missing entries by y_1 and y_4

Since only four values of f (x) are given , the polynomial which fits the data of degree 3.

Hence the fourth difference are zero .

$\Delta^4 y_k = 0$

$(E - 1)^4 (y_k) = 0$ .... (1)

Put k =0

$$(E^4 - 4E^3 + 6E^2 - 4E + 1)(y_0) = 0$$

$$y_4 - 4y_3 + 6y_2 - 4y_1 + y_0 = 0$$

$$y_4 - 4(15) + 6(8) - 4y_1 + 0 = 0$$

$y_4 - 4y_1 - 12 = 0$

$y_4 - 4y_1 = 12$........................ (2)

$put\ k = 1$

$(E^4 - 4E^3 + 6E^2 - 4E + 1)(y_1) = 0$

$Y_5 - 4y_4 + 6y_3 - 4y_2 + y_1 = 0$

$35 - 4y_1 + 90 - 32 + y_1 = 0$

$-4y_1 + y_1 = -93$............ (3)

Equation (2)× 4 $4y_4 - 16y_1 = 48$

$-4y_1 + y_1 = -93$

$-15y_1 = -45$

$y_1 = 3$

substituting $y_1 = 3$ in ....(1)

$y_4 - 4(3) = 12$

$y_4 = 24$

Hence, the missing entries are 3, 24.

13. Using Newton's forward interpolation formula find the cubic polynomial.

x	0	1	2	3
$f(x)$	1	2	1	10

Solution:

The forward interpolation formulae,

$Y(x = x0 + nh) = y0 + \frac{n}{1!}\Delta y0$ +n(n-1)/2

$\Delta^2 y_0 + \frac{n(n-1)(n-2)}{3!}\Delta^3 y_0 + \cdots$

Here, $x_o + nh = x \rightarrow x_o, h = 1$

0+n =x , n = x

The difference table is ,

X	Y= f(x)	$\Delta y0$	$\Delta^2 y_0$	$\Delta^3 y_0$
0	1			
1	2	1	-2	
2	1	-1	10	12
3	10	9		

$Y\ (n = h\ ) = \ y0 + \frac{n}{1!}\Delta y0$ +n(n-1)/2

$\Delta^2 y_0 + \frac{n(n-1)(n-2)}{3!}\Delta^3 y_0 + \cdots$

$$y = 1 + \frac{x}{1!}(1) + \frac{x(x-1)}{2!}(-2) + \frac{x(x-1)(x-2)}{6}(12)$$

$$y = 1 + x + (x^2 - x)(-1) + x(x^2 - 3x + 2)(2)$$

$$y = 1 + 6x - 7x^2 + 2x^3$$

Hence ,the cubic polnominal is

$2x^3 - 7x^2 + 6x + 1$

14. The population of a city in a censes taken once in 10 years is given below. Estimate the population in the year 1955

Year	1951	1961	1971	1981
Population in lakhs	35	42	58	84

Solution

$x_0 + nh = 1955 \rightarrow 1951 + n(10) = 1955$ $[h = 10]$

$10n = 1955 - 1951 = 4$

$$n = \frac{4}{10} = 0.4$$

The difference table is

X	Y= f(x)	$\Delta y0$	$\Delta^2 y_0$	$\Delta^3 y_0$
1951	35			
1961	42	7		
1971	58	16	9	
1981	84	26	10	1

The forward interpolation formulae,

$$Y(x = x0 + nh) = y0 + \frac{n}{1!}\Delta y0 + n(n-1)/2$$

$$\Delta^2 y_0 + \frac{n(n-1)(n-2)}{3!}\Delta^3 y_0 + \cdots$$

$$y_{(x=1955)} = 35 + 0.4(7) + \frac{(0.4)(0.4-1)}{2}(9) + \frac{(0.4)(0.4-1)(0.4-2)}{6}(1)$$

$$y = 35 + 2.8 - 1.08 + 0.064$$

$$y_{(x=1955)} = 36.784$$

Hence , the population in the year 1955 is 36.784

15. From the following table find the number of students who obtained marks less than

Marks	30-40	40-50	50-60	60-70	70-80
No. of Students	31	42	51	35	31

Solution

Let x be the marks and y be the number of students

By converting the given series into cumulative frequency distribution, the difference table is as follows.

X	Y= f(x)	$\Delta y0$	$\Delta^2 y_0$	$\Delta^3 y_0$	$\Delta^4 y_0$
Less than 40	31				
		42			
50	73		9		
		51		-25	
60	124		-16		37
		35		12	
70	159		-4		
		31			
80	190				

$$Y(x = x0 + nh) = y0 + \frac{n}{1!}\Delta y0 + n(n-1)/2$$

$$\Delta^2 y_0 + \frac{n(n-1)(n-2)}{3!}\Delta^3 y_0 + \cdots$$

Here, $x_o + nh = 45 \rightarrow x_o = 40, h = 10, n = \frac{1}{2}$

$$= 31 + 21 - \frac{9}{8} - \frac{25}{16} - 37 \times \frac{15}{128}$$

$$= 47.864$$

16. For the given data find the value of $\Delta^3 y_0$ is

x	5	6	9	11
y	12	13	15	18

Solution

$\Delta^3(y_0) = 0$

$$= (E-1)^3(y_0)$$

$=(E^4 - 4E^3 + 6E^2 - 4E + 1)\,(y_0)$

$= y_3 - 3y_2 - 3y_1 - y_0$

$= 18 - 45 + 39 - 12$

= 57-57

=0

17. The following data are taken from the steam table.

Tempreture C^0	140	150	160	170	180
Pressure kg f/cm^2	3.685	4.854	6.302	8.076	10.225

Find the pressure at temperature $t = 175^0$

Solution

Since the pressure required is at the end of the table, we apply **Backward interpolation formula.** Let temperature be x and the pressure be y.

$$Y(x = x0 + nh) = y0 + \frac{n}{1!}\nabla yn + n(n-1)/2$$

$$\nabla^2 y_0 + \frac{n(n-1)(n-2)}{3!}\nabla^3 y_0 + \cdots$$

$$x_o + nh = 175 \rightarrow x_o = 180, h = 10, n = -0.5$$

X	Y= f(x)	$\nabla y0$	$\nabla^2 y_0$	$\nabla^3 y_0$	$\nabla^4 y_0$
140	3.685				
		1.169			
150	4.854		0.279		
		1.4448		0.047	
160	6.032		0.326		0.002
		1.774		0.049	
170	8.076		0.375		
		2.149			
180	10.225				

$$y_{(x=175)} = 10.225 \pm 0.5(2.149) + \frac{(-0.5)(0-5)}{2!}(0.375)$$

$$+\frac{(-0.5)(0-5)(1.5)}{3!}\left(0.049 + \frac{(-0.5)(0-5)(1.5)(2.5)}{4!}\right)(0.002)$$

$$= 9.10048438$$

= 9.1

18. Using Lagrange's interpolation formula find y(10) from the following table:

x	5	6	9	11
y	12	13	14	16

Solution

Here the intervals are unequal. By Lagrange's interpolation formula we have

$X_0=5, x_1=6, x_2=9, x_3=11$

$y_0=5, y_1=6, y_2=9, y_3=11$

$$y = f(x) = \frac{(x-x_1)(x-x_2)(x-x_3)}{(x_0-x_1)(x_0-x_2)(x_0-x_3)} \times y_0 + \frac{(x-x_0)(x-x_2)(x-x_3)}{(x_1-x_0)(x_1-x_2)(x_1-x_3)} \times y_1$$
$$+ \frac{(x-x_0)(x-x_1)(x-x_3)}{(x_2-x_0)(x_2-x_1)(x_2-x_3)} \times y_2 + \frac{(x-x_0)(x-x_1)(x-x_2)}{(x_3-x_0)(x_3-x_1)(x_3-x_2)}$$

$$y(10) = f(10) = \left(\frac{4 * (1)(-1)}{(-1)(-4)(-6)}\right) * 12 + \left[\frac{5 * (1)(-1)}{(1)(-3)(-5)}\right]$$
$$\times 13 + \left[\frac{5 * (4)(-1)}{4 * (3)(-2)}\right] * 14 + \left[\frac{5(4)(1)}{6(5)(2)}\right] * 16$$

$$= \frac{12}{6} - \frac{13}{3} + \frac{5(14)}{6} + \frac{16(4)}{12}$$

$$= 14.663$$

19. Using interpolation, find the value of $f(x)$ when $x = 15$

x	3	7	11	19
$f(x)$	42	43	47	60

Solution

$X_0=3, x_1=7, x_2=11, x_3=19$

$y_0=42, y_1=43, y_2=47, y_3=60$

$:x= 15$

$$y = f(x) = \frac{(x-x_1)(x-x_2)(x-x_3)}{(x_0-x_1)(x_0-x_2)(x_0-x_3)} \times y_0 + \frac{(x-x_0)(x-x_2)(x-x_3)}{(x_1-x_0)(x_1-x_2)(x_1-x_3)} \times y_1 + \frac{(x-x_0)(x-x_1)(x-x_3)}{(x_2-x_0)(x_2-x_1)(x_2-x_3)} \times y_2 + \frac{(x-x_0)(x-x_1)(x-x_2)}{(x_3-x_0)(x_3-x_1)(x_3-x_2)}$$

$$= \frac{42}{4} - 43 + 12 \times \frac{47}{+15}$$

$$= 10.5 - 43 + 70.5 + 15$$

$$y = 53$$

X=15, f(x)=53

20. Using graphic method find the value of *y* when *x* = 48 from the following data:

x	40	50	60	70
y	6.2	7.2	9.1	12

Solution

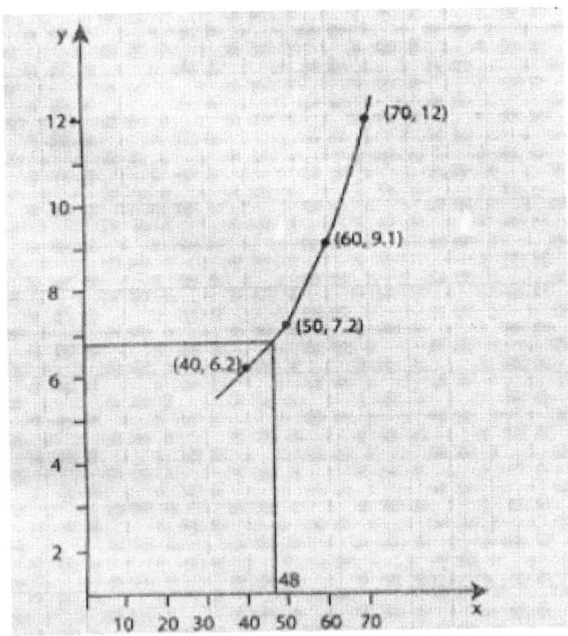

From the graph we can find x= 48 when the value of y is equal to 6.8

References

Sinha, K.C. (1992).A Text Book of Calculus, Student's (Publisher).

Aggarwal,R.S. and Aggarwal D.(2001). A comprehensive Guide for MCA Entrance Examination. Revised Edition, Sultan Chand & Sons, Delhi.

Gupta, S.C. and Kapoor, V.K. (2002). Fundamentals of Mathematical Statistics, Eleventh Revised Edition, Sultan Chand & Sons, Delhi.

9 789359 191546

Printed at Repro India Ltd.